Applied Science Review™

Botany

Applied Science Review™

Botany

Carl R. Pratt, PhD
Associate Professor of Biology
College of New Rochelle
New Rochelle, N.Y.

Springhouse Corporation
Springhouse, Pennsylvania

Staff

EXECUTIVE DIRECTOR, EDITORIAL
Stanley Loeb

PUBLISHER, TRADE AND TEXTBOOKS
Minnie B. Rose, RN, BSN, MEd

ART DIRECTOR
John Hubbard

CLINICAL CONSULTANT
Maryann Foley, RN, BSN

EDITORS
Diane Labus, David Moreau, Karen Zimmermann

COPY EDITORS
Diane M. Armento, Pamela Wingrod

DESIGNERS
Stephanie Peters (associate art director),
Matie Patterson (senior designer)

COVER ILLUSTRATION
Scott Thorn Barrows

ILLUSTRATORS
Jacalyn Facciolo, Jean Gardner, Judy Newhouse,
Stellarvisions

MANUFACTURING
Deborah Meiris (director), Anna Brindisi,
Kate Davis, T.A. Landis

EDITORIAL ASSISTANTS
Caroline Lemoine, Louise Quinn, Betsy K. Snyder

Library of Congress Cataloging-in-Publication Data
  Pratt, Carl Richard.
  Botany / Carl R. Pratt
    p. cm. – (Applied science review)
  Includes bibliographical references and index.
  1. Botany – Outlines, syllabi, etc. I. Title II. Series.
QK52.P73 1994
581 – dc20   93-10888
ISBN 0-87434-570-7   CIP

Contents

Advisory Board

Leonard V. Crowley, MD
    Pathologist
    Riverside Medical Center
    Minneapolis;
    Visiting Professor
    College of St. Catherine, St. Mary's
      Campus
    Minneapolis;
    Adjunct Professor
    Lakewood Community College
    White Bear Lake, Minn.;
    Clinical Assistant Professor of Laboratory
      Medicine and Pathology
    University of Minnesota Medical School
    Minneapolis

David Garrison, PhD
    Associate Professor of Physical Therapy
    College of Allied Health
    University of Oklahoma Health Sciences
      Center
    Oklahoma City

Charlotte A. Johnston, PhD, RRA
    Chairman, Department of Health
      Information Management
    School of Allied Health Sciences
    Medical College of Georgia
    Augusta

Mary Jean Rutherford, MEd, MT(ASCP)CS
    Program Director
    Medical Technology and Medical
      Technicians—AS Programs;
    Assistant Professor in Medical Technology
    Arkansas State University
    College of Nursing and Health Professions
    State University

Jay W. Wilborn, CLS, MEd
    Director, MLT-AD Program
    Garland County Community College
    Hot Springs, Ark.

Kenneth Zwolski, RN, MS, MA, EdD
    Associate Professor
    College of New Rochelle
    School of Nursing
    New Rochelle, N.Y.

Reviewers

Lorraine Mineo, BS, MA
    Lecturer
    Lafayette College
    Department of Biology
    Easton, Pa.

Leslie Ruth Towill, PhD
    Associate Professor of Botany
    Department of Botany
    Arizona State University
    Tempe

Dedication

To my wife Margie, my Mom, and my Dad.

Preface

This book is one in a series designed to help students learn and study scientific concepts and essential information covered in core science subjects. Each book offers a comprehensive overview of a scientific subject as taught at the college or university level and features numerous illustrations and charts to enhance learning and studying. Each chapter includes a list of objectives, a detailed outline covering a course topic, and assorted study activities. A glossary appears at the end of each book; terms that appear in the glossary are highlighted throughout the book in boldface italic type.

    Botany provides conceptual and factual information on the various topics covered in most botany courses and textbooks and focuses on helping students to understand:

- the molecular and cellular structure of plants
- the growth and development of plant roots, stems, leaves, flowers, fruits, and seeds
- the processes involved in cellular respiration and photosynthesis
- the control mechanisms by which plants thrive and interact with their environment
- the significance of soil and nutrition in plant growth
- the identification and classification of plants and plant-like organisms
- the importance of plant ecology and plant genetics.

1

Molecules, Cells, and Tissues

Objectives

After studying this chapter, the reader should be able to:
- Name the major classes of molecules and describe their functions in plants.
- Explain the difference between prokaryotic and eukaryotic cells.
- Describe the cellular events of mitosis and meiosis.
- Discuss the functions of the structures found in plant cells.
- Discuss the functions of meristematic and nonmeristematic plant tissues.

I. Molecular Composition of Cells

A. General information

1. Botany is a branch of biology that deals with the properties and life phenomena of plants; as with other living matter, the study of plant life usually begins with cellular composition
2. Although organisms are tremendously varied, all cells are remarkably similar in their basic structures, particularly at the molecular level
3. About 99% (by weight) of all living matter is composed of carbon, hydrogen, nitrogen, oxygen, phosphorus, and sulfur
4. Water makes up more than 50% of all living tissue and more than 90% of most plant tissues
5. There are 92 different types of naturally occurring atoms, but only a few are used to form the complex, highly organized molecules of living things
6. Molecules are groups of atoms joined together by chemical bonds
   a. Chemical bonds are links of pure energy that result from the sharing or complete transfer of electrons between atoms
   b. Bonds hold together similar or different types of atoms to form the many kinds of molecules
   c. Chemical bonds are classified according to the type of interaction between the atoms
      (1) Covalent bonds result from a sharing of electrons by two atoms
      (2) Ionic bonds result from the complete transfer of electrons from one atom to another; an atom either gains or loses electrons
7. All organic compounds contain carbon, and they may also contain hydrogen, oxygen, and nitrogen
   a. Carbon atoms tend to form covalent bonds rather than ionic bonds

b. Organic molecules, which are classified according to their chemical structure and properties, may consist of long chains of carbon atoms

8. The four classes of biologically important molecules are the carbohydrates, lipids, proteins, and nucleic acids
   a. Molecules belonging to these four classes are called ***macromolecules*** because of their large size
   b. Most macromolecules consist of smaller repeated units, or building blocks
9. ***Monomers*** are the building blocks of large molecules; many monomers link together to form large molecules known as ***polymers***
   a. Through the process of ***dehydration synthesis***, monomers are chemically joined by the removal of water to form polymers
   b. Through ***hydrolysis***, polymers are split chemically into their component monomers by the addition of water

B. Carbohydrates

1. Carbohydrates, the most abundant organic compounds in nature, are typically composed of carbon, hydrogen, and oxygen in a ratio of 1:2:1 (C:2 H:O)
2. Carbohydrates include sugars, starches, and related substances
3. They may be used as energy sources for the cell, such as energy storage units, or as structural components, such as membranes and ***organelles***
4. Carbohydrates, like other macromolecules, are classified according to their structure and function
   a. *Monosaccharides* are simple sugars
      (1) They usually consist of five or six carbon atoms that form a ring
      (2) Common ones include glucose, ribose, deoxyribose, and fructose
         (a) Glucose has six carbon atoms and the formula $C_6H_{12}O_6$
         (b) Ribose has five carbon atoms and the formula $C_5H_{10}O_5$
         (c) Deoxyribose also has five carbon atoms but lacks one oxygen atom; its formula is $C_5H_{10}O_4$
         (d) Fructose, or "fruit sugar," has six carbon atoms and a chemical formula identical to that of glucose
      (3) Many monosaccharides, such as glucose and fructose, have a short life span in the cell because they are either metabolized to free energy for use in cellular reactions or are linked together to form disaccharides or polysaccharides
   b. *Disaccharides* are two monomer units joined by dehydration synthesis
      (1) Common disaccharides include sucrose (table sugar), which consists of one unit of glucose joined to one unit of fructose, and maltose (malt sugar), which consists of two units of glucose
      (2) When the cell needs energy, disaccharides are commonly converted to monosaccharides by hydrolysis
      (3) In plants, disaccharides are the type of sugar that is commonly transported from one area to another
   c. ***Polysaccharides*** are long chains of monosaccharides used as energy storage molecules and structural components of cells
      (1) Starch, the chief storage polysaccharide in plants, consists of long chains of glucose molecules
         (a) It may occur as a coiled, unbranched chain of up to 1,000 glucose subunits

(b) It more commonly occurs as huge branched chains of up to 500,000 glucose subunits

(2) Glycogen, a common storage molecule in fungi and bacteria, is a polymer of glucose; it is usually smaller than starch, with small branches every 10 to 12 glucose subunits

(3) *Cellulose*, the principal structural polysaccharide in plants, is found in *cell walls*

    (a) It is the most abundant natural polysaccharide

    (b) Cellulose is composed of chains of glucose, as is starch and glycogen, but the chains are oriented differently (every other glucose subunit is upside down), thereby giving cellulose different biologic functions

(4) Chitin, a structural polysaccharide found in the cell walls of fungi, consists of a six-carbon sugar with many nitrogen groups attached

(5) Pectin, another important structural polysaccharide in plants, serves as a glue to hold adjacent cells together, particularly in fruits

C. Lipids

1. A diverse group of molecules that includes fatty and oily substances, lipids serve as energy storage units and waterproof coverings around cells

2. All lipids are hydrophobic (water fearing) and insoluble in water (and therefore often water repellent)

3. Lipids yield more energy per gram (9.3 kcal/g) than do carbohydrates (3.8 kcal/g) when used as an energy source by cells

4. Lipids include fats, cutin, *suberin*, waxes, and phospholipids

   a. Fats (saturated and unsaturated), also called *triglycerides*, consist of three fatty acids joined to a glycerol molecule

    (1) Saturated fats consist of carbon atoms that hold (are bonded to) as many hydrogen atoms as possible

      (a) They generally exist as animal fats, such as lard and butter

      (b) They melt at temperatures higher than those required to melt unsaturated and polyunsaturated fats

    (2) Unsaturated and polyunsaturated fats contain carbon atoms joined together by double bonds; the carbons can form additional bonds with various other atoms

      (a) Commonly produced by plants, unsaturated fats tend to be oils

      (b) Examples include safflower oil, peanut oil, and corn oil

   b. Cutin and suberin are lipid polymers that are found in the cell walls of certain plant tissues

    (1) Cutin and suberin form a matrix in which waxes (long-chain lipids) are imbedded; they differ from one another in the types of imbedded waxes and their chemical composition

    (2) *Cuticle* (cutin with embedded waxes) forms the outermost covering of the epidermal cells of leaves, flowers, and fruits and prevents excess water loss from these above-ground plant parts

    (3) Suberin is found in the cork cells of bark and the *Casparian strip* of root endodermal cells

   c. Phospholipids are closely related to fats

    (1) Each phospholipid consists of a glycerol molecule attached to two fatty acids and a phosphorus-containing molecule

(2) Phospholipids have a "split personality"
  (a) The fatty acid portion contains hydrocarbons, making the molecule hydrophobic (water repellent)
  (b) The phosphate portion is hydrophilic (water attracting), making the molecule water soluble
  (c) In water, phospholipids tend to form a film; the hydrophobic portion faces away from water and the hydrophilic portion dissolves in water
(3) They are the predominant component of cellular membranes

D. Proteins

1. **Proteins** are the principal structural and regulatory molecules of cells
2. These large polymers are composed of hundreds of **amino acids** (nitrogen-containing monomers)
   a. Every amino acid has the same fundamental structure, consisting of a central carbon bonded to an amino group ($-NH_2$), a carboxyl group ($-COOH$), a hydrogen, and a variable group represented by the letter R
   b. The different R-groups attached to the amino acid determine the specific type of amino acid
   c. **Peptide bonds** are formed when the amino group of one amino acid is linked to the carboxyl group of the next amino acid
3. The same basic set of 20 amino acids is found in nearly all proteins, but the different arrangements and proportions of these amino acids are responsible for the wide variety of proteins found in organisms
4. Proteins form highly organized shapes or conformations, which are extremely important to their function
   a. If a protein loses its shape, it loses its function **(denaturation)**
   b. If a protein regains its shape, it may regain its function **(renaturation)**
5. Biologists recognize four levels of protein organization
   a. The primary structure is the sequence of amino acids
   b. The secondary structure is the repeated, regular structure assumed by protein chains, which commonly form a helix or pleated sheet
   c. The tertiary structure is the complex three-dimensional structure of a single peptide chain
   d. The quaternary structure is the complex three-dimensional structure of a protein composed of more than one peptide chain
6. Proteins give structure and form to cells and regulate chemical reactions within cells
7. Enzymes are proteins that regulate chemical reactions within cells by altering the rate of the reactions

E. Nucleic acids

1. Nucleic acids are polymers of nucleotides
2. Each **nucleotide** consists of a phosphate group, a five-carbon (pentose) sugar, and a nitrogenous base (a ring-like structure that contains nitrogen and carbon atoms)
3. Two types of nucleic acid exist
   a. **Deoxyribonucleic acid** (DNA) has a double-helix structure

  (1) The double helix of DNA consists of two long chains of nucleotides that twist or spiral around an imaginary axis to form a molecular structure reminiscent of a spiral staircase or ladder

  (2) The two vertical members of the helix "ladder" are formed by the alternating sugar and phosphate groups of the nucleotides

  (3) The nitrogenous bases of the nucleotides are paired in the interior of the helix

  (4) DNA stores genetic information as a code represented by the linear sequence of the four different nitrogenous bases of the nucleotides; the pattern of the nucleotide bases determines, via the genetic code, different amino acids in protein synthesis

  (5) DNA is a self-replicating molecule that can faithfully ensure its own duplication

  (6) It contains deoxyribose sugar ($C_5H_{10}O_4$), a form of the pentose sugar that contains one less oxygen atom than ribose ($C_5H_{10}O_5$)

 b. ***Ribonucleic acid*** (RNA) appears in three forms within cells and is usually a single-stranded molecule containing ribose sugar

  (1) *Ribosomal RNA* (rRNA) is a component of the cellular structure that is responsible for synthesizing polypeptides known as ribosomes

  (2) *Messenger RNA* (mRNA) is a molecule that carries the information to make a specific polypeptide; it is synthesized from DNA and used by the ribosomes to help in the synthesis of polypeptide chains

  (3) *Transfer RNA* (tRNA) is a small RNA molecule which ensures that the proper amino acid is added to a growing peptide chain during protein synthesis

4. Each nucleotide of DNA or RNA contains two types of nitrogenous bases

 a. A pyrimidine base is a single-ring structure

  (1) The three pyrimidines are thymine, cytosine, and uracil

  (2) DNA contains thymine and cytosine; RNA contains cytosine and uracil

 b. A purine base is a double-ring structure

  (1) The two purines are adenine and guanine

  (2) DNA and RNA contain both types of purines

5. Nucleotides other than those used to synthesize nucleic acids include energy carrier molecules, such as adenosine triphosphate (ATP), which carry energy from place to place within a cell, and intracellular messengers, such as cyclic adenosine monophosphate (AMP), which coordinate chemical reactions within cells

II. Cell Structure and Function

A. General information

1. Cells are the basic unit of structure and function in all living things

2. They vary in diameter from about 0.2 to 50 microns or more (a micron is one-millionth of a meter or one-hundredth the thickness of a human hair)

3. Cells are classified according to the presence of an organized nucleus and internal components (organelles), such as endoplasmic reticulum, vacuoles, plastids, and mitochondria

      a. ***Prokaryotic cells*** contain no organized nucleus, nuclear membrane, or membranous organelles; bacteria and cyanobacteria (blue-green algae) are prokaryotic cells

        (1) Prokaryotic cells are small (usually less than 5 microns) and most have a cell wall

        (2) The DNA of prokaryotic cells is coiled and attached to the cell membrane to form a structure called a nucleoid; the DNA is not separated from the rest of the cell

      b. ***Eukaryotic cells*** contain an organized nucleus with a nuclear membrane and numerous membranous organelles; the cells of plants, animals, fungi, and protists (single-cell organisms) are eukaryotic cells

        (1) Eukaryotic cells are larger than prokaryotic cells (usually more than 10 microns)

        (2) The DNA of eukaryotic cells lies within the nucleus, where it is separated from the rest of the cell

4. In addition to internal structures, all cells possess an outer envelope known as a cell or plasma membrane; some cells (such as plant cells) possess an additional external structure known as a cell wall

B. Cell wall

1. The cell wall is a rigid or semirigid structure that surrounds the membrane of some cells
2. Cell walls are present in plants, bacteria, and fungi, but not in animal cells
3. The principal component of plant cell walls is cellulose; cell walls also contain lignin and lipids

      a. Lignin adds strength, rigidity, and disease resistance

      b. Lipids (cutin, suberin, and waxes) help the cell control the movement of water

4. The cell wall varies in thickness and is commonly layered to strengthen and protect the cell

      a. All plant cells have a ***primary cell wall***; this wall is found in actively dividing cells and mature cells involved in metabolic processes, such as photosynthesis, respiration, and secretion

      b. The middle lamella, which is composed of pectin, is located between the primary walls of adjacent cells

      c. Many other cells develop a ***secondary cell wall***

        (1) The secondary cell wall forms inside the primary cell wall after the cell stops growing

        (2) The secondary wall strengthens the cell

      d. Adjacent cells are held together by the polysaccharide pectin, which is found in the middle lamella

      e. ***Plasmodesmata*** are tiny passageways in cell walls through which cytoplasm from one cell extends into adjacent cells, providing a pathway for the movement of fluids and dissolved substances from cell to cell

C. Cytoplasm

1. The ***cytoplasm*** consists of the cellular fluid within the cell (cytosol) and the cell organelles
2. ***Cytosol*** is largely made up of water and various dissolved substances, such as proteins, amino acids, lipids, and carbohydrates

D. Cell membrane
1. The *plasma* or *cell membrane* is the outer boundary of the cell
2. It is composed of phospholipids and proteins with some carbohydrates attached to the outer surface
3. The cell membrane forms a semipermeable barrier that controls substance movement in and out of the cell

E. Organelles of plant cells
1. Plant and animal cells share many structures, such as the endoplasmic reticulum, ribosomes, mitochondria, and Golgi apparati; however, plants contain a number of structures not found in animals, such as a cell wall, plastids, and vacuoles
2. The *endoplasmic reticulum* is an internal cell membrane that forms channels and pockets in the cytoplasm; it partitions the cytoplasm, acts as an attachment site for enzymes and ribosomes, and forms passageways through the cytoplasm
3. *Ribosomes* are small globular structures that serve as a site of protein synthesis and may be attached to the endoplasmic reticulum or unattached within the cytoplasm; they are found in all prokaryotic and eukaryotic cells
4. *Mitochondria* are the powerhouses of plant and animal cells
   a. These numerous, tiny, typically sausage-shaped structures are 1 or more microns long
   b. They have an outer membrane and an inner membrane, which is folded to form cristae
   c. The cristae are the site of most of the cell's ATP production
      (1) ATP serves as an energy carrier
      (2) It is the cellular fuel for nearly all cellular activities
   d. Mitochondria tend to accumulate in groups where energy (ATP) is needed
5. *Golgi apparati*, or dictyosomes, are clusters of flat, roundish sacs
   a. Golgi apparati are the site of polysaccharide synthesis and chemical alteration of proteins and phospholipids
   b. Material produced or altered within the Golgi apparati are enveloped by vesicles for transport from the cell (exocytosis)
6. *Plastids*, a dynamic group of self-regulating, nearly autonomous plant cell organelles, comprise chloroplasts, chromoplasts, and leucoplasts
   a. *Chloroplasts* are spindle-shaped plastids approximately 5 microns long
      (1) They convert solar energy (sunlight) to a chemical form (carbohydrate) through a process known as photosynthesis
      (2) Chloroplasts contain *stroma*, *thylakoids*, and *grana*
         (a) Stroma is a semiliquid material in which carbon fixation occurs during photosynthesis
         (b) Thylakoids are flattened sacs of membranes that contain *chlorophyll* and other accessory pigments used to collect and harvest light energy
         (c) Grana are stacks of thylakoids
   b. *Chromoplasts* are diversely shaped plastids that are approximately the same size as chloroplasts; they contain various pigments (usually yellow, orange, or red) and are responsible for giving plant parts their color
   c. *Leucoplasts* are also variously shaped and similar in size to chloroplasts

        (1) These colorless plastids synthesize and store various materials within the cell
        (2) Two types of leucoplasts exist
            (a) **Amyloplasts** synthesize and store starch
            (b) **Elaioplasts** synthesize and store oils
    d. Plastids can convert from one type to another
        (1) For example, the chloroplasts that make an unripened tomato appear green gradually convert to chromoplasts that contain red pigments, making the now ripened fruit appear red
        (2) If exposed to light, some leucoplasts may develop into chloroplasts
7. **Vacuoles** and **vesicles** are membrane-bounded regions of the cytoplasm
    a. Vacuoles are large liquid-filled cavities found only in plants; they store water and water-soluble substances
    b. Vesicles are small structures found in both plant and animal cells; they are involved in the transport of materials to and from the cell membrane

F. Nucleus and nuclear membrane
1. Commonly the most prominent structure in the cytoplasm of eukaryotic cells, the nucleus averages 5 microns in diameter
2. Its two primary roles are to control ongoing cellular functions by determining which protein molecules are synthesized and to store and transmit genetic information
3. The nucleus is enclosed by a **nuclear membrane**, which regulates the movement of materials into and out of the nucleus
4. When the cell is not dividing, the DNA and associated proteins within the nucleus appear as a grainy fluid (**chromatin**) dispersed throughout the nucleus
5. The nucleus also may contain one or more dark-staining regions known as nucleoli (**nucleolus**, singular), which are the sites of ribosomal RNA synthesis

III. Cellular Reproduction

A. General information
1. Cell division of prokaryotic cells (those without a nucleus) differs from that of eukaryotic cells (those with a nucleus)
2. Prokaryotic cells reproduce by **binary fission**
    a. The original (parent) cell divides to produce two identical daughter cells, each of which receives a copy of the single parental chromosome
    b. When the daughter cells reach a certain size, they also divide
3. Eukaryotic cells reproduce by **mitosis**, a process of cell division characterized by sequential and equal allocation of chromosomes to both daughter cells
4. In sexual reproduction, the reproductive cells are produced by a cell division process called **meiosis**

B. Cell cycle
1. The cell cycle is the sequence of events that occurs when cells divide
2. The duration of the cell cycle depends on the type of cell and external factors, such as temperature and nutrient availability
3. The two phases of the cell cycle are mitosis and interphase
    a. *Mitosis* is the period of cell division

   b. *Interphase* is the time between cell divisions; it consists of three periods
      (1) During G_1 (gap period 1), which occurs after mitosis, cytoplasmic material grows
      (2) During S (synthesis period), the genetic material (DNA) in each chromosome is duplicated
      (3) During G_2 (gap period 2), the proteins that will be used to construct the spindle fibers during mitosis are synthesized
      (4) During all three periods, the genetic material (**chromosomes**) is in a dispersed, diffuse state known as chromatin

C. Mitosis
   1. Mitosis results in the division of a parent cell into two genetically identical daughter cells
   2. It involves two related events known as karyokinesis and cytokinesis
      a. **Karyokinesis** is the division of chromosomes (collections of DNA and protein) in the nucleus
      b. **Cytokinesis** is the division of the cytoplasm, parceling of the organelles, and separation of the parent cell into daughter cells; in plants, cytokinesis occurs by formation of a cell plate
   3. Mitosis has four stages—prophase, metaphase, anaphase, and telophase
      a. *Prophase* is the first stage
         (1) During this phase, the chromosomes condense to form short, thick fibers
            (a) When viewed through a light microscope, these fibers appear as X-shaped structures
            (b) Each half of the "X" is called a **chromatid**, and the two chromatids (called a chromatid pair) are attached in the center by a **centromere**
         (2) The nuclear membrane begins to disappear
         (3) Spindle fibers begin to form from proteins synthesized earlier during G_2 of interphase
         (4) **Centrioles** (found in animals, fungi, some algae, and some ferns, but not in seed plants) replicate and move toward each end, or pole, of the cell
         (5) Chromatids begin to migrate toward the center, or equator, of the cell
      b. *Metaphase* is the second stage
         (1) Chromatids align across the equator of the cell
         (2) Spindle fibers attach to the centromeres of the chromatids
         (3) The nuclear membrane disappears completely
         (4) Centrioles, if present, arrive at the poles of the cell
      c. *Anaphase* is the third stage
         (1) The centromeres of each chromatid pair divide and separate (each with one chromatid attached)
         (2) The separate chromatids are drawn toward opposite poles of the cell by contraction of the attached spindle fibers
      d. *Telophase* is the fourth and final stage of mitosis
         (1) Chromatids arrive at opposite poles of the cell
         (2) Spindle fibers disappear
         (3) Nuclear membranes form around each bundle of chromatids
   4. The separation of cytoplasm follows nuclear division

a. Cytokinesis usually begins in telophase and may be completed during the early portion of interphase (G_1)

b. In animal cells, cytokinesis occurs through the formation of a cleavage furrow, which cuts across the cell equator; further constriction at the equator divides the parent cell into two daughter cells

c. In plant cells, a cell plate forms from a series of vesicles derived from the Golgi apparatus; the vesicles collect at the cell equator and gradually fuse, thereby separating the daughter cells

d. In some algae and fungi, mitosis occurs without subsequent cytokinesis, resulting in *coenocytic* cells with many nuclei but without separate membranes and walls

D. Meiosis

1. Meiosis, which occurs within the reproductive organs of sexually reproducing organisms, results in the formation of gametes (in animals) or spores (in plants)

   a. Each gamete is *haploid;* that is, it contains just one set of chromosomes

   b. When the gametes (egg and sperm) fuse during *fertilization*, a *zygote* is formed; the zygote is *diploid;* that is, it contains two complete sets of chromosomes, one from each parent

      (1) The two sets of chromosomes, called *homologous chromosomes*, are comparable in size and shape

      (2) They bear similar, but not necessarily identical, information about genetic traits or characteristics

   c. Through mitosis, cells divide and the zygote grows, but it continues to have a diploid number of chromosomes within each body cell

   d. Through meiosis, the diploid number of chromosomes within the gametes is halved, resulting in haploid cells

2. Meiosis occurs in two major stages—meiosis I and meiosis II

   a. *Meiosis I* reduces the number of chromosomes in half

      (1) Prophase I is the first stage of meiosis I

         (a) The duplicated chromosomes found as chromatin during interphase condense to reveal their chromatid form

         (b) The nuclear membrane begins to disappear

         (c) Chromosomes pair with their homologues (mates), forming homologous chromosomes

         (d) Homologous pairs wrap around each other in a process called *synapsis* and commonly exchange genetic material, called *crossing over*

         (e) Centrioles, if present, migrate to the poles of the cell

         (f) Spindle fibers begin to appear

      (2) Metaphase I is the second stage of meiosis I

         (a) Homologous pairs move to the cell equator

         (b) Centromeres of each homologous pair attach to spindle fibers extending from opposite poles of the cell

      (3) Anaphase I is the third stage of meiosis I

         (a) Spindle fibers pull each homologous pair away from its homologous partner and toward opposite poles of the cell

         (b) Centromeres do not divide at this point, as they do in mitosis and meiosis II

      (4) Telophase I is the final stage of meiosis I

(a) Paired chromosomes arrive at opposite poles of the cell and usually disperse into chromatin form

(b) In some cells, the nuclear membrane temporarily reforms around the chromatids; in other cells, no nuclear membrane forms, and the cells immediately move into meiosis II

b. *Interkinesis*, or *interphase II*, is a period of rest between the first and second meiotic divisions

(1) Interkinesis is not universally present; some cells proceed immediately from telophase I to prophase II

(2) With or without interkinesis, genetic material does not replicate before the second meiotic division

c. *Meiosis II* separates the chromatids and results in the formation of four haploid cells

(1) Prophase II is the first stage of meiosis II

(a) Chromatin condenses to form chromatids

(b) Spindle fibers begin to form

(c) Centrioles, if present, migrate to opposite poles

(2) Metaphase II is the second stage of meiosis II

(a) Spindle fibers form completely

(b) Chromatid pairs migrate to the cell equator

(c) Spindle fibers attach to the centromeres of chromatid pairs

(d) Centromeres of each chromatid pair divide, and chromatids separate as they do in mitosis

(3) Anaphase II is the third stage of meiosis II

(a) Single chromatids (each half of the original chromatid pair) are pulled to opposite sides of the cell equator

(b) Chromatids are pulled toward the poles by contraction of the spindle fibers

(4) Telophase II is the final stage of meiosis II

(a) Four haploid cells are formed, each containing only half of the genetic information of the parent cell

(b) A nuclear membrane forms around each new nucleus, and the chromatids disperse to chromatin form

(c) Cytokinesis is initiated and may be completed by the end of telophase II or may continue into the beginning of the interphase

IV. Plant Tissues

A. General information

1. Tissues are collections of cells that function as a unit

2. Plants contain *meristematic tissues* (in which the cells actively divide) and *nonmeristematic tissues* (in which the cells no longer divide)

B. Meristematic tissues

1. **Apical meristems**, the centers of cell division in actively growing plants, are the ultimate source of all cells in the plant body

a. These tissues are found at the tips of roots and shoots

b. They add new cells by mitosis, which increases the length or height of these structures

2. **Primary meristems** develop from each apical meristem
    a. The three types of primary meristems produce only **primary tissues** (see *Primary Meristems and Tissue*)
    b. **Protoderm** gives rise to the epidermis
    c. **Ground meristem** produces the bulk of the **stem** or root
    d. **Procambium** gives rise to vascular or transport tissue
3. *Lateral meristems*, which include vascular **cambium** and cork cambium, produce **secondary tissues**
    a. **Vascular cambium** is found in the stems and roots of perennial plants and some herbaceous annuals
        (1) It produces secondary vascular or conducting tissue
        (2) It increases the girth or diameter of the plant
    b. **Cork cambium** is found in the stems and roots of certain plants, especially woody varieties
        (1) It produces the outer bark of trees
        (2) It also produces the secondary tissue, called periderm, that replaces the epidermis of roots and stems

C. Nonmeristematic tissues

1. Simple nonmeristematic tissues are composed of essentially one cell type, such as parenchyma, collenchyma, or sclerenchyma
    a. **Parenchyma** cells typically possess only a primary cell wall, although some have a secondary cell wall
        (1) Parenchyma cells are the most abundant cells in nonmeristematic tissues
        (2) They play a role in metabolic processes, such as photosynthesis, cellular respiration, and food storage, and are the progenitors of all other tissue types
        (3) They also may transport food and water
        (4) These cells are found in the cortex of stems and roots, the pith of stems, the leaf mesophyll, and the flesh of fruits; some intercellular air spaces exist between the cells
        (5) Although considered nonmeristematic tissue, parenchyma cells can be stimulated to divide throughout their life span; they do so in order to initiate regeneration and wound healing if the plant is damaged
    b. **Collenchyma** cells have only primary cell walls
        (1) The thick walls of collenchyma cells provide strength to the plant
        (2) Collenchyma cells generally support young growing organs; they continue to develop thick, flexible walls while the organ in which they are found continues to grow
        (3) These cells are found in strands or layers in stems and petioles (leaf stalks) and around the vascular tissue of leaves
        (4) Collenchyma cells are alive at maturity
    c. **Sclerenchyma** cells have thick secondary walls often containing **lignin**
        (1) The two types of sclerenchyma cells are sclereids and fibers
            (a) Sclereids are variable-shaped cells; they exist in the stone cells (cells that give certain fruits a gritty texture) of pears and other fruits, seed coats, and nut shells
           (b) Fibers are long, slender cells that are organized in strands or bundles, such as those of hemp, cotton, and linen

Primary Meristems and Tissue

As a young plant undergoes primary growth, apical meristems give rise to three primary meristems: protoderm, ground meristem, and procambium. The primary meristems in turn produce the variety of primary tissues that make up the plant body.

Protoderm differentiates into covering tissues, including the epidermis of roots, stems, and leaves. More specialized examples include guard cells, leaf hairs, and root hairs.

Young stem

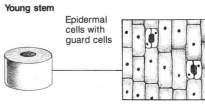

Epidermal cells with guard cells

Leaf

Root tip

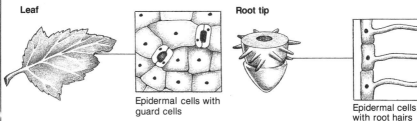

Epidermal cells with guard cells

Epidermal cells with root hairs

Ground meristem differentiates into three basic tissue types:
* Parenchyma is widely distributed in the stem and root and makes up the photosynthetic tissues of the leaf. These large, thin-walled cells are commonly involved in storage.
* Collenchyma is primarily involved in support. Its thick-walled cells form tough strands of tissue below the epidermis, within vascular tissue, and in the supporting portions of the leaf.
* Sclerenchyma, in its fiber form, strengthens young shoots. In its sclereid form, it provides hardness for seed coverings and shells.

Parenchyma

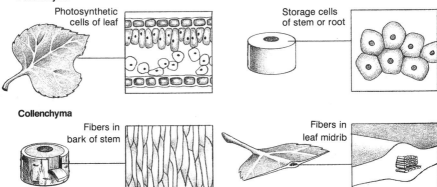

Photosynthetic cells of leaf

Storage cells of stem or root

Collenchyma

Fibers in bark of stem

Fibers in leaf midrib

Sclerenchyma

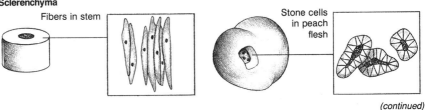

Fibers in stem

Stone cells in peach flesh

(continued)

Primary Meristems and Tissue *(continued)*

Procambium differentiates into xylem and phloem, the vascular tissue of roots, shoots, and leaves. Xylem transports water and minerals; phloem transports food.

Phloem   Xylem

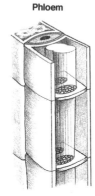
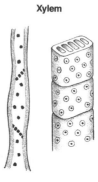

---

      (2) Sclerenchyma cells are commonly, though not always, dead at maturity

2. Complex nonmeristematic tissues are composed of several cell types

   a. The *epidermis* is the outermost layer of cells of the primary plant body

      (1) It may consist of parenchyma cells, guard cells, trichomes (hairs), and sclerenchyma cells

      (2) The epidermis usually is coated with a waxy surface layer called the cuticle

   b. *Xylem*, which may be a primary or secondary tissue, is the principal water-conducting tissue in plants

      (1) Xylem moves water and dissolved substances from root to shoot through water-conducting cells, which are dead, hollow, and lacking in cytoplasm

      (2) Two types of conducting cells exist—tracheids and vessels

         (a) *Tracheids* are found in most seedless vascular plants (ferns) and gymnosperms (conifers, pines, spruces, and hemlocks); water moves from cell to cell through perforations in the walls of the tapered, overlapping cell ends

         (b) *Vessels* are found (along with tracheids) in the xylem of angiosperms (flowering plants); water moves from cell to cell through tubelike cells joined end to end

   c. *Phloem*, which may be a primary or secondary tissue, is the principal food-conducting tissue in plants

      (1) It is composed of sieve tube cells and companion cells

         (a) Each sieve tube cell has pores (plasmodesmata) through which materials enter and exit the cell

         (b) Companion cells assist in the movement of substances into and out of the sieve tube cells

      (2) Unlike the passive transport in xylem cells, transport in phloem cells is active, which requires expenditure of energy by the plant

      (3) Because energy is required in phloem transport, phloem cells must be alive when they transport materials

d. The **periderm** (outer bark), a secondary tissue, commonly replaces the epidermis of stems and roots
   (1) It consists of cork tissue, cork cambium, and phelloderm and can last the lifetime of the plant
   (2) **Cork** is the predominant constituent of the periderm
       (a) This protective tissue dies shortly after forming, but it continues to protect the plant against water loss and mechanical damage
       (b) The cell walls of cork cells contain large amounts of suberin, a lipid that retards water loss from the stem
   (3) *Cork cambium* is a layer of actively dividing cells that produces cork tissue on the outside and phelloderm on the inside
   (4) **Phelloderm** is a layer of living parenchyma tissue that can replace damaged cork cambium

D. Tissue systems

1. All cells and tissues within the body of a plant are organized into three basic systems—dermal, vascular, and ground
2. Tissue systems are found in stems, roots, and leaves
3. They are initiated in the embryo from the three primary meristems
   a. Procambium produces the cells of the vascular tissue system
   b. Protoderm produces the cells of the dermal tissue system
   c. Ground meristem produces the cells of the ground tissue system
4. The *vascular tissue system* consists of xylem and phloem and conducts materials throughout the plant body
5. The *dermal tissue system,* or epidermis, is usually a single layer of cells that covers and protects the young parts of a plant
   a. The characteristics of dermal tissue vary with its function and location
   b. The dermal tissue of leaves secretes a waxy coating (cuticle) that reduces water loss
   c. The dermal tissue of roots produces cellular extensions (root hairs) to increase the absorptive surface for the uptake of water and minerals from soil
   d. The dermal tissue may be replaced by periderm (bark) later in plant growth
6. The *ground tissue system* fills the space between the dermal and vascular tissue systems
   a. The ground tissue is composed predominantly of parenchyma cells, with some collenchyma and sclerenchyma cells
   b. Ground tissue functions in photosynthesis, storage and transport, and adds to the bulk of the plant body

Study Activities

1. Describe the four classes of molecules found in plants.
2. Identify the organelles of plant cells and discuss their functions.
3. Summarize the steps of mitosis and meiosis.
4. List the different meristematic and nonmeristematic tissues and describe their functions.

2

Stems, Roots, and Leaves

Objectives

After studying this chapter, the reader should be able to:
- Describe the basic structure of a plant.
- Explain the role of the primary meristems in plant growth.
- Compare and contrast the stem structures in monocots and dicots.
- Give examples of specialized stem structures.
- Describe the regions of a growing root.
- Sketch a cross section of a typical leaf and describe its features.
- Explain why leaves change color in autumn.
- Describe the process of abscission.

I. Basic Plant Body

A. General information
    1. Plants are composed of individual cells, which are organized into tissues and then arranged to form organs
    2. The plant body develops from apical *meristems*, which in turn develop into primary meristems (which produce primary tissues) and secondary meristems (which produce secondary tissues)
    3. The plant body is divided into above-ground structures (*shoots*) and below-ground structures (roots)

B. Shoots
    1. Shoots arise from the *plumule* (which consists of *epicotyl* and *leaf primordium*) during embryonic development
    2. Shoots can be divided into functional parts, such as *stems,* leaves, flowers, and fruits
       a. Stems are the support structures of the plant and can store food
       b. Leaves are the site of photosynthesis
       c. Flowers and fruits play a role in reproduction and seed production and dispersal

C. Roots
    1. In most vascular plants, roots make up the underground portion of the plant
    2. Roots anchor the plant and extract water and nutrients from soil

II. Stems

A. General information
1. Stems make up the bulk of the above-ground portion of a plant
2. They are commonly divided into small stem portions called branches
3. Stems display a variety of forms
    a. Most stems have an erect form of branches and leaves (such as those seen in roses, tulips, and trees)
    b. Some stems grow horizontally near or beneath the soil surface (such as those in ferns and some grasses)

B. Functions
1. Stems produce leaves and flowers and hold them upright
2. They contain vascular tissue, which conducts water, nutrients, and food throughout the plant body
3. They may serve as storage areas for sugar and starch produced by photosynthesis

C. External structure
1. The external features of the stem include the node, internode, buds, and leaf scar
2. A *node* is the point at which a leaf or branch attaches to a main stem
3. The *internode* is the part of the stem between successive nodes
4. *Buds* are small embryonic regions that are commonly protected by young leaves
    a. *Axillary buds* arise where leaves are attached to the stem and may develop into branches or flowers
    b. *Terminal buds*, located at the tips of branches, grow to extend the length of the branch or stem
        (1) Terminal bud scale scars are tiny scars that mark the point of attachment of the previous year's terminal bud
        (2) The distance between terminal bud scale scars represents one year's stem growth
    c. *Lateral buds* are found adjacent to terminal buds; they may develop into branches or flowers
5. The *leaf scar* marks the location of a leaf before it fell from the plant

D. Growth and development
1. The stages of stem growth and development can be observed by moving from the terminal bud toward the base of the stem
    a. Younger, immature tissues are found near the terminal bud
    b. Older, mature, and differentiated tissues are located farther from the terminal bud
2. Growth begins with cell division at the apical meristem and continues within the three primary meristems (protoderm, ground meristem, and procambium)
    a. *Protoderm* gives rise to the epidermis (external protective layers of the stem)
    b. *Ground meristem* gives rise to the inner portions of the stem, the central core or *pith*, and the surrounding layers of cells known as the *cortex*
        (1) The cortex is a primary tissue comprised largely of parenchyma cells
        (2) In plants with extensive secondary growth, the cortex is obliterated
    c. *Procambium* gives rise to xylem and phloem (vascular and transport tissue)

Comparison of Three Stem Cross Sections

In the *herbaceous dicot* stem, the vascular bundles, which contain primary xylem and phloem, are arranged around the perimeter of the stem. In the *monocot* stem, the vascular bundles are distributed throughout the stem. The *woody dicot* stem contains an abundance of secondary xylem, forming wood inside a stem; vascular cambium separates the xylem from the secondary phloem. The bark includes all tissues outside of the vascular cambium.

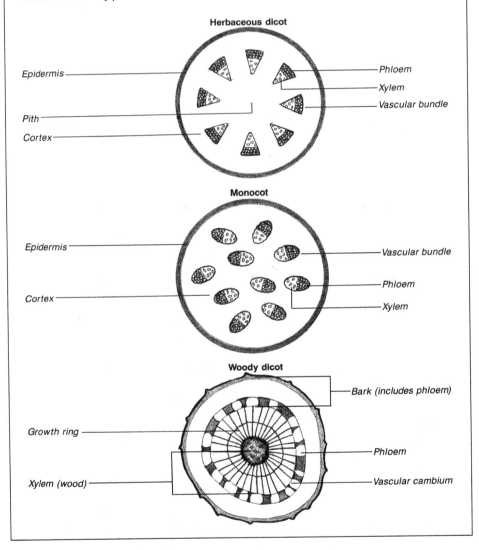

3. Depending on the type and species of plant, growth and development may be limited to primary tissues produced by the apical and primary meristems
4. In some plants, the lateral meristems (vascular cambium and cork cambium) produce further growth in diameter or girth
    a. *Vascular cambium* gives rise to secondary vascular tissues, which add girth to the stem

(1) It develops from the procambium as a narrow band of cells

(2) Vascular cambium is found between the primary xylem and phloem and retains its meristematic ability to divide

b. *Cork cambium* produces cork cells, which contain suberin, a fatty substance that makes the cells impervious to water

(1) Cork protects the stem from water loss and mechanical damage

(2) Cork cambium develops in the outer regions of the stem of certain plants, especially woody varieties

E. Tissue arrangement

1. Tissue is arranged in different patterns within the stems of different plant types (see *Comparison of Three Stem Cross Sections*)

a. **Dicotyledon**, or dicot, plants have two cotyledons (food storage organs) in their seeds

b. **Monocotyledon**, or monocot, plants have only one storage organ

2. Stems are classified as herbaceous (soft, pliable) or woody (hard, treelike, and containing secondary xylem)

3. Many *herbaceous dicots*, such as alfalfa and sunflowers, are **annuals** (plants that complete their life cycle in one growing season)

a. Herbaceous species consist mostly of primary tissue and generally do not have a vascular cambium

b. The primary xylem and primary phloem are arranged in discrete patches (**vascular bundles**) that are located in a ringlike pattern between the cortex and pith

4. *Woody dicots*, such as trees and shrubs, are usually long-lived and often display considerable secondary growth

a. Woody species contain a vascular cambium, which produces secondary xylem and secondary phloem

(1) Secondary xylem, also called wood, is concentrated in concentric rings within the middle of the stem

(2) Secondary phloem is pushed to the outer edges of the stem

(3) The vascular cambium produces more secondary xylem than secondary phloem

(4) The volume of secondary xylem and phloem produced by the vascular cambium generally obliterates the primary vascular tissue, although in some plants small amounts of primary xylem and pith may remain in the center of the stem

b. As the stem widens in diameter, the increased girth crushes the secondary phloem, and the vascular cambium produces rays (transverse or cross-sectional lines of specialized parenchyma cells) to transport water and nutrients laterally in the stem

c. As the stem matures, the xylem organizes into two regions known as heartwood and sapwood

(1) **Heartwood** consists of the inner xylem, which is filled with resins, gums, and tannin; it is darker than sapwood because of its accumulation of waste materials, and it no longer functions in conduction

(2) **Sapwood** consists of the xylem actively involved in the transport of water and dissolved substances; it constitutes the relatively new secondary xylem near the vascular cambium

(3) Heartwood and sapwood give wood its characteristic color patterns

    d. The outermost region of the woody stem is called **bark**

       (1) It encompasses all tissues outside the vascular cambium, including phelloderm, phellogen, cork, cork cambium, and phloem

       (2) Bark protects the tree from water loss, provides some thermal insulation, protects against mechanical abrasion and some insect damage, and functions in food transport (because it contains phloem tissue)

    e. Cambial growth in woody dicots is affected by the temperature and the amount of water and light available

       (1) Spring wood is produced early in the growing season when a flush of growth occurs; xylem cells are large and generally consist of xylem vessels

       (2) Summer wood is produced as growth slows later in the season; cells are smaller and primarily consist of tracheids

       (3) The annual cycling between spring and summer wood creates the annual growth rings of trees; these rings not only tell the age of a tree but also provide information about growth conditions

5. *Monocots* are generally herbaceous and have only primary tissues

    a. They do not contain a vascular or cork cambium

    b. The xylem and phloem are found in discrete vascular bundles scattered throughout the cross section of the stem

    c. Some monocots (wheat, rice, barley, oats, rye, and other similar grasses) contain an **intercalary meristem** (region of cell division found in the node rather than the apical meristem) at the base of each internode

       (1) The intercalary meristem increases the length of the stem but does not appreciably change the stem's diameter

       (2) It allows grass to grow back after cropping or cutting

F. Specialized stems

1. **Rhizomes** are horizontal stems that grow below or near the ground's surface

    a. Externally, they resemble roots and can be mistaken for roots because of their location

    b. Common in ferns and many grasses, rhizomes often contain parenchyma tissue, which can store starch

    c. Rhizomes are important in vegetative reproduction because they allow the plant to spread laterally

2. **Stolons** are horizontal stems, or "runners," that grow above ground and often develop buds, which can later develop into new plants (such as the stolons of strawberry plants and spider plants)

3. Tubers, **bulbs**, and corms are highly modified underground stems that function as food storage organs

    a. **Tubers** are short, enlarged, fleshy stems that often swell from food storage; the potato is a tuber

    b. *Bulbs* are short stems covered by fleshy leaves filled with stored food; they are characteristic of onions, lilies, and tulips

    c. **Corms** are similar to bulbs, but are flattened and consist of stem tissue surrounded by papery scale-like leaves; examples include crocus and gladiolus

III. Roots

A. General information
1. Roots are the first portion of the plant to emerge from the seed; the emerging root is called the *radicle*
2. Roots share many structural and developmental characteristics with stems
3. As with stems, the stages of root growth and development can be seen by moving away from the root tip toward the base
4. Monocots generally do not have secondary tissues in roots, just as they do not have secondary tissues in stems; monocot roots also have a centrally located pith, which is absent in dicot roots

B. Functions
1. Roots absorb and store water and mineral nutrients from the soil
2. They also anchor or hold the plant in place
3. Because they possess meristematic tissues, roots may become active in reproduction if they are severed or separated from the plant

C. Structure
1. The growing root can be divided into several regions; each region has specific characteristics associated with development
2. The *root cap* covers the root tip and protects the growing root from damage as it pushes through the soil
3. The *region of cell division* is surrounded and protected by the root cap
   - a. This region contains apical meristem, which lies just under the root cap and is functionally equivalent to the apical meristem of stems
   - b. The apical meristem (in roots as well as in stems) is divided into three meristematic regions
      - (1) Protoderm gives rise to the epidermis, or outer layer of cells
      - (2) Ground meristem produces the cortex
      - (3) Procambium produces vascular tissue (xylem and phloem)
4. The *region of elongation*, located immediately behind the region of cell division, extends approximately 0.5" (1 cm) from the root tip
   - a. In this region, cells rapidly increase in size, but not in number, which increases the length of the root
   - b. Cells increase in size by absorbing large amounts of water
5. The *region of maturation*, which is adjacent to the region of elongation, is where root cells and tissues fully mature and differentiate
   - a. This region is characterized by cellular differentiation (a process by which cells acquire specialized structures to perform specific functions)
   - b. *Root hairs* develop in the region of maturation
      - (1) These cellular projections of the epidermis form the outermost layer of the growing root
      - (2) Root hairs are numerous and greatly increase the surface area, and thus the absorptive area, of the root
      - (3) In a growing root, the extent of the root hair zone remains fairly constant, with new root hairs forming (within the region of maturation) toward the root cap and old root hairs dying in the more mature regions; the average root hair lives only a few days

Cross Section of a Root

The central portion of the root (vascular cylinder or stele) is surrounded by the endodermis. The individual endodermal cells possess bands known as Casparian strips, which allow the cells to control the contents of the vascular cylinder.

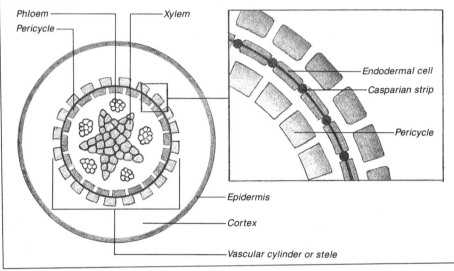

c. The *cortex* also develops in the region of maturation and persists as the plant matures; however, in plants with substantial secondary growth, the cortex disappears as bark is produced

d. The **endodermis**, a single layer of specialized cells, marks the inner boundary of the cortex (see *Cross Section of a Root*)
 (1) The endodermis grows to surround the vascular tissue located in the center of the root
 (2) Each endodermal cell produces a suberin band around its perimeter, known as a *Casparian strip*
 (a) Casparian strips block the passage of water through the otherwise permeable cell walls of the cortex
 (b) Because of the Casparian strips, all water and materials dissolved in water must pass through endodermal cells to enter the xylem; the endodermal cells act as "gateway" cells, controlling the contents of the vascular tissue

e. The **stele**, or central portion of the root, consists of the pericycle, xylem, and phloem
 (1) The **pericycle**, a single layer of cells inside the endodermis, produces lateral or branch roots and helps the endodermis move materials into the stele
 (2) Primary xylem forms the star-shaped core of the stele
 (3) Phloem tissue lies between the "arms" of the xylem
 (4) In woody plants, the cork cambium of the root develops from the pericycle and gives rise to cork tissue (periderm)

D. Specialized roots

1. All roots have some capacity for food storage
2. Plants that live in arid regions may have roots with cortex cells that contain large vacuoles for water storage
3. Plants that grow in swampy areas, where their roots are likely to be underwater, have specialized roots called **pneumatophores**, which deliver air to other roots
4. Many plants produce buds along roots that grow near the surface of the ground; these buds can develop into small plants known as suckers
5. Corn, mangroves, and banyan trees commonly produce additional roots where the stem enters the ground; these roots are called prop roots because they provide additional support to the plant

IV. Leaves

A. General information

1. Leaves are considered part of the shoot system of a plant
2. As the most recognizable portion of a plant, leaves show considerable variety in size, shape, arrangement, and form

B. Functions

1. In most plants, leaves are the principal site of photosynthesis
2. Water evaporates from leaves when they are exposed to sunlight; this process is called transpiration
3. Leaves can become highly modified and function in food storage
4. If leaves become modified to form spines, they can serve a protective role

C. External structure

1. Leaves consist of a blade and a petiole
   a. The **blade**, the flattened and commonly widened portion of the leaf, functions in photosynthesis
   b. The **petiole**, the stalklike portion of the leaf, is attached to a branch or stem
2. Leaves are attached to stems in one of three arrangements
   a. *Alternate leaves* are those in which a single petiole is attached to a node
   b. *Opposite leaves* are those in which two petioles are attached to a node on opposite sides of the stem
   c. *Whorled leaves* are those in which three or more petioles are attached to a node, often forming a circle around the stem
3. Leaves are classified as simple or compound
   a. *Simple leaves* are those in which each blade is attached by a single petiole to the stem
   b. **Compound leaves** are those in which the leaf is subdivided into smaller entities, or leaflets, and the leaflets are attached to the stem by a common petiole
      (1) In pinnately compound leaves, the leaflets occur in pairs along a central stalklike structure called a *rachis*
      (2) In palmately compound leaves, the leaflets are attached at the same point at the end of the petiole
4. Leaves differ in their *venation* (arrangement of veins or vascular tissue)

Cross Section of a Leaf

A typical leaf consists of upper and lower layers of epidermis, with the palisade mesophyll and spongy mesophyll sandwiched in between. Photosynthesis occurs primarily in the palisade mesophyll.

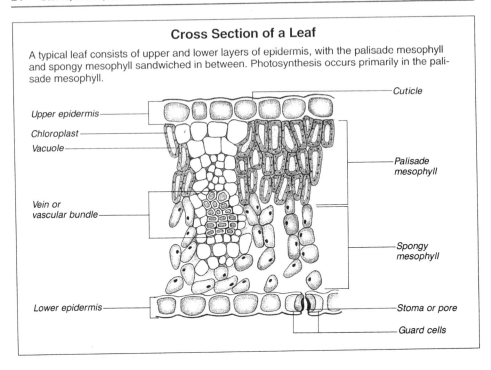

a. *Parallel venation* occurs when leaf veins are oriented more or less parallel to the leaf edge (margin) and to each other

b. *Pinnate venation* occurs when a main vein or midrib courses through the center of the leaf with secondary veins branching from it

c. *Palmate venation* occurs when several veins diverge from the base of the leaf and spread to the margins

d. Parallel venation is characteristic of monocots; pinnate and palmate venation are characteristic of dicots

D. Internal structure

1. The leaf consists of several layers (see *Cross Section of a Leaf*)

2. The *epidermis* is the outer layer of cells located on the upper and lower surface of the leaf

   a. Its coating of wax (cuticle) retards water loss from leaf surfaces

   b. The epidermis contains tiny pores called stomata (***stoma***, singular) that permit gas exchange

   c. Stomata openings are controlled by adjacent guard cells, which can adjust to light and moisture conditions

3. The ***palisade mesophyll*** is a layer of columnar cells beneath the epidermis

   a. Palisade cells contain a large central vacuole surrounded by numerous chloroplasts

   b. The palisade mesophyll is the principal site of photosynthesis

4. The ***spongy mesophyll***, located beneath the palisade mesophyll, is a layer of irregularly shaped cells with large intercellular spaces

   a. This layer makes up much of the thickness of the leaf

   b. Spongy mesophyll cells, like palisade cells, contain many chloroplasts

   c. The numerous intercellular spaces contained in the spongy mesophyll are connected to the outer atmosphere through the stomata; the intercellular spaces facilitate the rapid gas and water vapor exchange that is crucial to efficient photosynthesis

5. The *leaf veins* are the vascular tissues (xylem and phloem) of a leaf
   a. The vascular tissues in leaves are surrounded by a collection of cells known as **bundle sheath cells**
   b. Leaf veins carry water and nutrients to the photosynthetic tissues and transport the products of photosynthesis (sugars) to other plant parts
   c. They also provide strength to the leaf

E. Specialized leaves

1. Leaves become specialized to adapt to environmental conditions and to perform specific functions
2. Leaves adapt to the intensity of sun exposure
   a. Leaves shaded by other leaves of the same plant or by leaves from different plants tend to be thinner than those exposed to full sunlight
   b. Shaded leaves have fewer epidermal hairs and fewer chloroplasts
   c. They also tend to have larger cells, but fewer well-defined mesophyll layers
3. Leaves adapt to the amount of moisture in the environment
   a. Aquatic environments, such as ponds, lakes, and streams, are considered *hydric*
      (1) Leaves adapted to hydric environments may float on the water surface or be submerged
      (2) Leaves that float (water lilies and pondweed) have large air chambers in the mesophyll for buoyancy; a thick cuticle prevents excess water from entering the leaf, and the stomata are located on the upper surface of the leaf to allow gas exchange without water infiltration
      (3) Submerged leaves (elodea and water milfoil) have no cuticle and lack stomata; they tend to be thin to allow gas exchange directly through the epidermis
   b. Terrestrial environments with moderate amounts of moisture are considered *mesic*
      (1) Leaves adapted to mesic environments have well-developed palisade and spongy mesophyll layers
      (2) Typical mesic environment plants include most trees (maple, oak, ash, birch), common vegetables (onion, corn, tomato), and garden flowers (tulip, rose, chrysanthemum)
   c. Dry environments are considered *xeric*
      (1) Leaves adapted to xeric environments tend to have thick cuticles and may have several layers of epidermis; the stomata are commonly indented within pits
      (2) The leaves also may be curled, which reduces surface exposure to the sun and wind and curtails evaporation
      (3) Typical xeric environment plants include laurel, rhododendron, and creosote bush
      (4) In severely xeric conditions, the leaves are modified for water storage or are absent; in leafless plants, such as cacti, the stem becomes photosynthetic
4. Leaves adapt for protection
   a. **Spines** are modified leaves used largely for protection from herbivores

b. Spines are typically found in desert plants and some woody plants (black locust, acacia)
5. Leaves adapt to provide food storage
a. They may become modified to contain large numbers of thin-walled parenchyma cells that swell with stored food and become fleshy
b. The fleshy leaves of the onion bulb are an example of this type of adaptation
6. Leaves are modified for reproduction
a. In some species (such as Kalanchoe), small plantlets develop along the serrated leaf margin
b. These plantlets drop to the ground and establish a new plant
7. Leaves may be altered to appear as flower petals (**bracts**), as seen in poinsettia flowers
8. Leaves can adapt to obtain nutrients
a. In carnivorous plants, complex leaf structures attract and capture insects as a method of obtaining sufficient nutrients, particularly nitrogen
b. Examples include the Venus fly trap, with a rapidly closing leaf trap; sundew, with sticky hairs on the leaf to trap insects; bladderwort, with a trap that captures prey; and pitcher plant and cobra lily, which have large tubular leaves filled with liquid to trap and digest prey

F. Color changes in autumn
1. A plant's color changes in autumn result from *pigments*
a. Pigments are molecules that absorb specific wavelengths of light
b. They are responsible for producing the various colors of leaves and other plant parts
2. During the growing season, chlorophyll, a green pigment, is present in concentrations sufficient to mask the presence of other pigments
3. In autumn, as the day shortens and temperatures fall, trees begin to metabolize chlorophyll and remove it from leaf tissues; as chlorophyll is removed, the other leaf pigments are revealed
4. Different plant species have different pigments in the leaf tissue
a. Typical pigments include chlorophyll (green), carotenes (yellow), xanthophyll (pale yellow), anthocyanin (red if the cell's cytoplasm is acidic, blue if alkaline, and intermediate shades if neutral), and betacyanins (red)
b. Because different tree species have different amounts of pigments in their leaves, they display distinctive colors in the fall; examples include aspen (yellow), maples (reds, oranges, and golds), ash (yellow), and sumac (red)

G. Abscission
1. **Abscission** is the shedding of leaves and other plant parts, such as flowers or fruits, as a result of environmental or physiologic changes
2. The **abscission zone** is a layer of cells located near the base of the petiole; these cells play a role in separating the plant organ from the plant body
3. During the growing season, *auxin* (a plant growth hormone produced in the leaf blade) inhibits the growth of the **abscission initials**, a small cluster of dormant parenchyma cells
4. As the leaf ages and the daylight and temperature conditions change in autumn, the production of auxin by the leaf blade decreases
a. The abscission initials then begin to grow and effectively wall off the petiole from the stem

b. Eventually, all that holds the leaf in place are a few strands of xylem, which can be easily broken by wind or rain

Study Activities

1. Sketch and label a cross section of a monocot stem, herbaceous dicot stem, and woody dicot stem.
2. Discuss the structure and function of rhizomes, stolons, tubers, bulbs, and corms, and give examples of each.
3. Sketch a growing root tip and describe the function of each region.
4. Define epidermis, endodermis, pericycle, and bundle sheath.
5. Describe the various leaf arrangements and venation patterns.
6. List at least three pigments found in leaves and the colors they produce.

3

Flowers, Fruits, and Seeds

Objectives

After studying this chapter, the reader should be able to:
- List the parts of a typical flower and describe their functions.
- Name the various types of fruits and give examples of each.
- Identify the parts of a seed and describe their functions.
- Explain the value of seed dormancy.
- List the steps of seed germination.

I. Flowers

A. General information
1. Flowers are the sexual reproductive organs of *angiosperms* (flowering plants)
2. They often function to attract *pollinators* (organisms that transfer pollen from one flower to another)
3. Not all plants have flowers; plants such as mosses, ferns, and conifers have a variety of reproductive structures

B. Structure
1. A flower consists of several parts (see *Flower Structures* for an illustration)
   a. The **peduncle**, or pedicel, is a short branch that the supports the entire flower
   b. The **receptacle** is the point at which the flower parts attach to the peduncle
   c. The **sepal** is a small, often green, leaflike bract that covers and protects the flower in the bud stage; collectively, the sepals are referred to as the *calyx*
   d. The **petals** are often colorful and showy structures, which generally serve to attract pollinators; collectively, the petals are referred to as the **corolla**
2. The male reproductive structures of a flower consist of one to many **stamens**
   a. The stamen comprises an anther and a filament
      (1) The **anther** is located at the top of the stamen; it contains numerous sacs that produce, store, and release **pollen** (male reproductive cells)
      (2) The thin, slender filament supports the anther
   b. In most flowers, the release of pollen is facilitated by lengthwise slits that form on the anthers
3. The female reproductive structures of a flower consist of one to many vase-shaped pistils, each of which comprises a stigma, style, and ovary

Flower Structures

Three types of flowers are depicted in this illustration: hypogynous (superior ovary), in which flower parts are attached to the receptacle beneath the ovary; perigynous, in which flower parts are attached to a cuplike structure that surrounds the ovary; and epigynous (inferior ovary), in which flower parts are attached to the receptacle above the ovary.

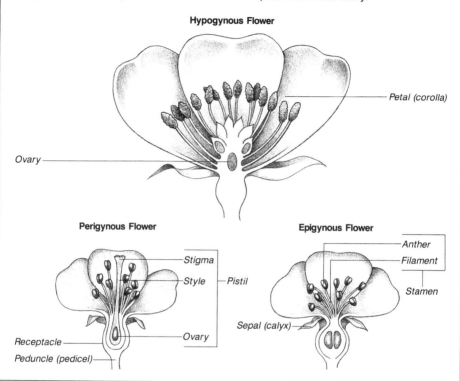

a. The *stigma*, which forms the top portion of the pistil, serves as the pollen receptor
b. The *style* is a long, slender filamentous structure that resembles the neck of a vase
    (1) It connects the stigma to the ovary
    (2) Upon germination, the pollen tube grows through the style toward the ovary
c. The *ovary* is located at the base of the pistil
    (1) It produces the eggs and associated structures
    (2) One or many ovules (female reproductive cells) are located within the ovary
    (3) Later in the life cycle, the ovules may develop into seeds and the ovary wall may develop into fruit

C. Types
1. The various arrangements of flowers are important in identifying plant species
2. Clusters or groups of flowers are called an ***inflorescence***

3. Flowers can be classified according to the position of the ovary (see *Flower Structures*, page 29)
   a. **Hypogynous** flowers **(superior ovary)** are those in which the sepals, petals, and stamens are attached to the receptacle beneath the ovary
   b. **Perigynous** flowers **(immediate ovary)** are those in which the sepals, petals, and stamens are attached to the side of the ovary
   c. **Epigynous** flowers **(inferior ovary)** are those in which the sepals, petals, and stamens are attached to the top of the ovary
4. Flowers are often classified according to the presence or absence of floral parts
   a. **Complete flowers** possess all flower structures (sepals, petals, stamens, and pistils)
   b. **Incomplete flowers** are missing one or more flower structures
   c. **Perfect flowers** contain both male (stamen) and female (pistil) reproductive structures
   d. **Imperfect flowers** contain only male or female reproductive structures
      (1) Those with only male structures are called **staminate**
      (2) Those with only female structures are called **carpellate** or **pistillate**

II. Fruits

A. General information
1. Fruits develop from the wall of an ovary (and occasionally other structures) and provide a means for dispersing seeds
2. Vegetables are plant parts other than fruits that are harvested for food; examples include stems (potatoes and celery), leaves (cabbage and lettuce), and roots (sweet potatoes and turnips)
3. Fruits that are often mistaken for vegetables include tomatoes, string beans, cucumbers, and squashes

B. Structure
1. **Pericarp** is the fruit wall that develops from the mature ovary wall; it can be divided into three regions
   a. The **exocarp** is the outer surface, or skin, of the fruit
   b. The **endocarp** is the inner region that surrounds the seed; it may become hard, as in a peach pit
   c. The **mesocarp** is the middle region located between the exocarp and endocarp; in fleshy fruits, mesocarp often makes up the bulk of the fruit
2. The three regions of the pericarp are present in the mature fruit, but may be difficult to distinguish from one another in some fruits

C. Types
1. Some fruits consist only of the seed and ovary; others may develop from or include adjacent flower parts
2. Fruits are classified by consistency or texture
3. *Fleshy fruits* have juicy or fleshy walls
   a. *Simple fleshy fruits* develop from flowers with a single pistil
      (1) The ovary may be superior or inferior; simple fruits may also develop from flower parts other than the ovary

      (2) A ***drupe*** contains a single seed enclosed by a hard stony endocarp, or pit (as in cherries, plums, olives, and apricots); the pericarp may be fleshy or fibrous or may wither and dry after the seed matures (as in pecans, pistachios, and walnuts)

      (3) A ***berry*** develops from a compound ovary and usually contains more than one seed; the pericarp is fleshy and it is difficult to differentiate mesocarp from endocarp

         (a) True berries have a thin skin and a fleshy pericarp at maturity; examples include tomatoes, grapes, persimmons, peppers, eggplants, avocados, and dates

         (b) Other berries develop from inferior ovaries and combine other flower parts to form the fruit; examples include blueberries, cranberries, and bananas

      (4) A *pepo* is a modified berry with an inseparable rigid rind composed of exocarp and flesh composed of mesocarp and endocarp; examples include pumpkins, cucumbers, watermelons, squash, and cantaloupes

      (5) A *hesperidium* is a modified berry with a separable leathery rind composed of exocarp and mesocarp; the endocarp is fleshy and composed of numerous outgrowths of the inner lining of the ovary wall

         (a) The outgrowths become saclike and swollen with juice as the fruit develops

         (b) Examples include oranges, lemons, limes, and grapefruits

      (6) A *pome* develops from an enlarged receptacle that grows around the ovary; its endocarp is leathery or papery

         (a) Examples include pears and apples

         (b) The core of an apple is the ovary and the rest of fruit is the receptacle

  b. ***Aggregate fleshy fruits*** develop from a single flower with several pistils

    (1) The individual pistils develop into tiny drupes and mature as a cluster on a single receptacle

    (2) Examples include raspberries, blackberries, and strawberries

  c. ***Multiple fleshy fruits*** develop from several individual flowers of a single inflorescence

    (1) Each flower has its own receptacle but matures and fuses to form a single larger fruit

    (2) Examples include mulberries, Osage oranges, pineapples, and figs

4. *Dry fruits* have a dry mesocarp

  a. *Dehiscent dry fruits* split open at maturity to release seeds

    (1) A ***follicle*** splits only along one side to expose seeds; examples include larkspurs, columbines, peonies, and milkweeds

    (2) A ***legume*** splits along two sides to release seeds; examples include peas, beans, lentils, and peanuts

    (3) A *silique* splits along two sides or seams, but the seeds are borne on a central partition, which is exposed at maturity; examples include broccoli, cabbage, radish, and watercress

    (4) In a ***capsule***, the ovary is divided into two or more compartments (carpels) that open at maturity; examples include irises, orchids, lilies, poppies, violets, snapdragons, and horse chestnuts

  b. *Indehiscent dry fruits* do not split open at maturity to release seeds

(1) An *achene* contains a single seed attached only to the base of the surrounding pericarp, which is relatively easy to separate from the seed; examples include sunflowers, buttercups, and dandelions

(2) A *nut* is a one-seeded fruit that is similar to, but generally larger than, an achene and has a thicker, harder pericarp; examples include acorns, hickory nuts, chestnuts, and filberts (hazelnuts)

(3) A *grain* (caryopsis) is also similar to an achene, but the pericarp is tightly fused to the seeds and cannot be separated; examples include corn, wheat, rice, oat, and barley

(4) A *samara* has a pericarp that surrounds the seed and extends outward to form a wing; examples include the fruits of maples, ashes, elms, and birches

III. Seeds

A. General information
1. **Seeds** function in the propagation of the plant
2. They are formed by the maturation of the ovule after fertilization

B. Structure
1. The seeds of monocot and dicot plants both consist of a seed coat and embryo
2. The **seed coat** is the outer layer, which develops from the integument (layers of tissue that surround the egg and are produced within the pistil)
3. The **embryo**, sometimes called the embryonic axis, is a tiny rudimentary plant with several identifiable parts
   a. The **epicotyl** lies just above the point of attachment for the cotyledons; it contributes to the growth of a stem, but may not be visible in the embryo
   b. The **plumule** develops into the shoot; it consists of the epicotyl and young, embryonic leaves
   c. The **cotyledon**, or "seed leaf," is attached just below the plumules
      (1) Monocotyledon (monocot) plants, such as corn, have only one cotyledon (also called the scutellum)
      (2) Dicotyledon (dicot) plants, such as beans, have two cotyledons
   d. The **hypocotyl** lies just below the point of attachment for the cotyledons
   e. The **radicle** is the portion of the embryo from which the primary root develops
      (1) In monocot seeds, the radicle is enclosed within a tubular sheath called the coleorhiza, and the plumule is enclosed within a tubular sheath called the coleoptile
      (2) In dicot seeds, the plumule and radicle are not enclosed within protective sheaths

C. Dormancy
1. Dormancy is a period of growth inactivity before germination
2. This adaptation allows the seed to survive harsh or unfavorable environmental conditions
3. Seeds may emerge from dormancy because of mechanical, environmental, or physiologic changes

a. **Scarification** can break dormancy through mechanical abrasion or scraping of the seed surface naturally or from human activity
b. **Stratification** entails four to six weeks of exposure to wet conditions and cold temperatures; it can be a natural or artificial method of breaking dormancy
c. Exposure to light can break dormancy in some seeds, while others must be buried in soil and hidden from light

D. Germination

1. Seed germination depends on exposure to appropriate soil nutrients, moisture conditions, daylight, and temperature—all of which may be unique to a particular type of seed
2. Seeds must absorb, or imbibe, water to initiate germination
3. Upon imbibition, the seed often generates considerable internal pressure that ruptures the seed coat and permits the radicle to emerge
4. During the early stages of germination, cellular activity increases dramatically, and the cells need considerable oxygen to germinate

Study Activities

1. Sketch a flower, label its parts, and explain their functions.
2. Describe the structures of simple, aggregate, and multiple fleshy fruits and give examples of each.
3. Describe the structures of dehiscent and indehiscent dry fruits.
4. List the parts of a seed and explain their functions.
5. Name three conditions that can break seed dormancy.

4

Cell Respiration and Photosynthesis

Objectives

After studying this chapter, the reader should be able to:
- Distinguish between heterotrophs and autotrophs.
- Describe the importance of adenosine triphosphate.
- List the steps of cellular respiration, fermentation, and photosynthesis.
- Explain the relationship between glycolysis and the Krebs cycle.
- Differentiate between aerobic and anaerobic respiration.
- Name the structures involved in photosynthesis.
- Discuss the importance of the Calvin-Benson cycle, Hatch-Slack pathway, Crassulacean acid metabolism, and photorespiration.

I. Sources of Cellular Energy

A. General information
    1. All living things require energy to perform activities necessary for survival
    2. All cells use *adenosine triphosphate* (ATP) as an energy source
        a. The ATP molecule is closely related to adenosine diphosphate (ADP)
            (1) Both molecules consist of a nitrogenous base (adenine) joined to ribose sugar and two (in the case of ADP) or three (in the case of ATP) phosphate groups
            (2) Addition of a third phosphate group to ADP results in ATP formation
        b. Cells store energy in the chemical bonds of the ATP molecule
        c. When the third phosphate group is chemically removed from ATP, energy (7.3 kcal/mole of ATP) is released
        d. Cells readily convert ADP to ATP and vice versa
            (1) Converting ADP to ATP enables the cell to store energy
            (2) Converting ATP to ADP enables the cell to release energy to fuel other chemical reactions within the cell
    3. Cells harvest energy from sunlight or from the organic molecules of other organisms
        a. *Autotrophs* are organisms, such as plants, that can convert the energy of sunlight into chemical energy
        b. *Heterotrophs* are organisms, such as animals, that cannot convert sunlight into chemical energy; they must harvest energy from organic molecules (sugar, proteins, lipids) formed by other organisms

B. Enzymes

1. *Enzymes* are protein catalysts that control the many chemical reactions of cells
2. Enzymes often require "helper" substances, known as cofactors and coenzymes, to control the chemical reactions
   a. *Cofactors* are ions that interact with enzymes and are essential for enzyme action; examples include magnesium ion, sodium ion, and potassium ion
   b. *Coenzymes* are nonprotein organic cofactors that play important roles in the enzyme-controlled reactions of photosynthesis and cellular respiration; examples include nicotinamide-adenine dinucleotide (NAD), flavin adenine dinucleotide (FAD), and nicotinamide-adenine dinucleotide phosphate (NADP)
      (1) NAD can donate electrons (oxidation) and accept electrons (reduction); it occurs in an oxidized form (NAD^+; the plus sign denotes the positive charge) and a reduced form (NADH; the H denotes the addition of electrons and a hydrogen atom)
      (2) FAD occurs in an oxidized form (FAD) and a reduced form ($FADH_2$)
      (3) NADP also has an oxidized form ($NADP^+$) and a reduced form (NADPH)
      (4) Oxidization and reduction are critical to the transfer of electrons in cellular chemical reactions

C. Adenosine triphosphate

1. When ATP is converted to ADP, the phosphate bonds release energy (an *exergonic* reaction)
2. When ADP is converted to ATP, the formation of the phosphate bonds absorbs energy (an *endergonic* reaction)
3. When ADP is converted to adenosine monophosphate (AMP) by the removal of the second phosphate group, additional energy is released
4. To prevent this loss of energy, cellular enzymes catalyze the transfer of the phosphate group from the ATP molecule to some other molecule
   a. The transfer of a phosphate group is called *phosphorylation*
   b. ATP and ADP are continuously recycled within a cell, transferring phosphate groups and the energy contained in their chemical bonds

II. Cellular Respiration

A. General information

1. *Cellular respiration* is a metabolic process by which energy is released from the breakdown of fuel (food) molecules
2. The chemical reactions of cellular respiration break down high-energy molecules (such as glucose), thereby resulting in the formation of energy-poorer molecules and the conversion of ATP to ADP
3. There are two types of cellular respiration
   a. *Anaerobic respiration* takes place in the absence of oxygen
   b. *Aerobic respiration* takes place in the presence of oxygen; it occurs in three stages: glycolysis, the Krebs cycle, and the electron transport chain
      (1) *Glycolysis* is a series of chemical reactions that break down glucose into pyruvate; this process does not require oxygen and takes place in the cytoplasm

(2) The ***Krebs cycle*** completes the breakdown of glucose, resulting in the release of energy and carbon dioxide; this process requires oxygen and takes place in the mitochondria

(3) The *electron transport chain* harvests additional energy from the electrons received from the Krebs cycle; this process, which requires oxygen and takes place in the mitochondria, forms most of the cell's ATP

4. The two methods by which ATP is formed during cellular respiration are substrate level phosphorylation and chemiosmosis

a. *Substrate level phosphorylation* is the direct enzyme-controlled transfer of a phosphate group to ADP from some other molecule

(1) Substrate phosphorylation occurs during glycolysis and the Krebs cycle

(2) Only a small percentage of the ATP produced during cellular respiration is formed this way

b. ***Chemiosmosis*** occurs when electrons obtained from the oxidation of the glucose molecule are used to produce ATP

(1) Most of the ATP produced during cellular respiration is formed this way

(2) Chemiosmosis occurs in three steps: removal of electrons from glucose or glucose fragments, operation of a proton pump, and ATP synthesis

5. Oxidation and reduction reactions are a major feature of cellular respiration

a. ***Oxidation*** (loss of electrons or hydrogens) and ***reduction*** (gain of electrons or hydrogens) are coupled reactions—one molecule loses when another molecule gains

b. Because oxidation and reduction occur simultaneously, the coupled reactions are called ***redox*** (reduction/oxidation) ***reactions***

B. Glycolysis

1. Glycolysis is the metabolic pathway whereby glucose ($C_6H_{12}O_6$) is split into two smaller fragments, which are then oxidized and rearranged to produce two pyruvic acid ($C_3H_4O_3$) molecules (see *Cellular Respiration*)

2. Two molecules are required to initiate the chemical reactions of glycolysis

a. Four ATP molecules are formed during glycolysis for each glucose molecule that is involved in this process, resulting in a net gain of two ATP molecules

b. Two NADH molecules are also formed; these molecules are later used to form ATP via ***oxidative phosphorylation***

c. The two pyruvic acid molecules are passed to the Krebs cycle

C. Krebs cycle

1. The Krebs cycle is linked to glycolysis by a transition reaction

a. The transition reaction results in the formation of acetyl coenzyme A (acetyl-CoA) by the oxidation (removal of carbon and oxygen atoms) of pyruvic acid; this oxidation reaction forms carbon dioxide and a 2-carbon intermediate

b. Electrons are also removed from pyruvic acid during the transition reaction, and NAD^+ is reduced to NADH

c. Coenzyme A, a derivative of vitamin A, attaches to the 2-carbon intermediate to form acetyl-CoA, which enters the Krebs cycle

2. The Krebs cycle is characterized by three events

a. Four carbon dioxide molecules are formed for every glucose molecule that enters the glycolysis pathway

Cellular Respiration

Glycolysis is linked to the Krebs cycle by the formation of acetyl coenzyme A from coenzyme A and pyruvic acid. The electron transport chain produces adenosine triphosphate by chemiosmosis.

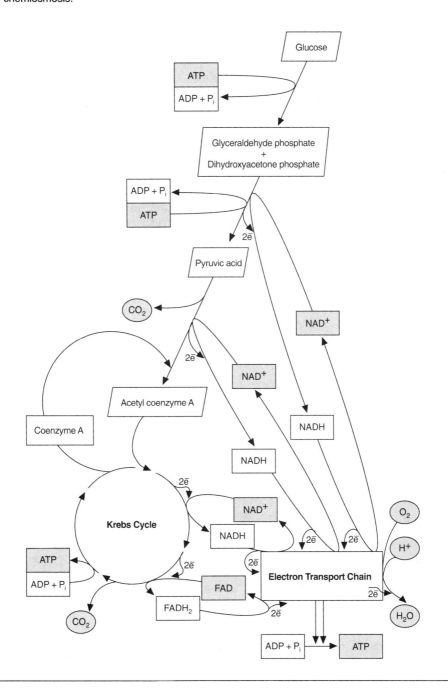

b. Six NADH and two $FADH_2$ molecules are formed from the oxidation of the glucose molecule

c. Two ATP molecules are produced by substrate phosphorylation

d. The reduced coenzymes NADH and $FADH_2$ carry electrons to an electron transport chain, where ATP is produced by chemiosmosis

D. Electron transport chain

1. The electron transport chain completes the oxidation of the glucose molecule that entered the glycolysis pathway

   a. It consists of a series of molecules (called electron carriers) that can readily accept and pass electrons to one another

   b. In the process of electron transfer, these molecules release energy

2. Electron carriers are embedded in the inner mitochondrial membrane

3. The transfer of electrons along the chain produces an accumulation of protons on the outside of the inner mitochondrial membrane

4. As the protons diffuse back through the inner membrane, ATP synthase, which is present in the mitochondrial membrane, phosphorylates ADP; this process is called chemiosmosis

   a. When NADH transfers electrons to the electron transport chain, it is oxidized to form NAD^+; the transferred electrons trigger the phosphorylation of three ADP molecules to three ATP molecules

   b. When $FADH_2$ transfers electrons to the electron transport chain, it is oxidized to form FAD; the transferred electrons trigger the phosphorylation of two ADP molecules to two ATP molecules

5. After the electron transport chain harvests energy from the electrons, the electrons are chemically combined with oxygen and protons to form water

6. The complete oxidation of a glucose molecule in aerobic respiration produces a total of 36 ATP molecules in eukaryotic cells and 38 ATP molecules in some bacterial cells

E. Fermentation

1. **Fermentation** is a form of anaerobic respiration in which organic nutrients are broken down to produce ATP

2. It consists of glycolysis and chemical reactions that regenerate NAD^+ by reducing pyruvic acid

3. The two most common types of fermentation are alcohol production and lactic acid formation

   a. *Alcohol production* occurs in many bacteria and yeast cells and some plant cells; it results in the conversion of pyruvic acid to ethanol

      (1) Carbon and oxygen atoms are removed from pyruvic acid to form carbon dioxide; the remaining 2-carbon compound is ethanol (C_2H_5OH)

      (2) NADH is oxidized to NAD^+ so that the glycolysis reactions can continue

   b. *Lactic acid formation* occurs in some eukaryotic cells (such as vertebrate muscle and liver cells) and some bacteria cells under anaerobic conditions

      (1) Pyruvic acid is reduced to lactic acid

      (2) NADH is oxidized to NAD^+

F. Alternative fuels for respiration

1. Although glucose is the primary fuel for cellular respiration, both aerobic and anaerobic respiration can make use of other fuels

2. Alternative fuels for respiration include polysaccharides, proteins, and fats
    a. Polysaccharides are hydrolyzed to monosaccharides that undergo glycolysis
    b. Proteins are hydrolyzed to amino acids, which are generally converted to pyruvic acid, acetyl-CoA, or other acids that are used in the Krebs cycle
    c. Fats are hydrolyzed to their component fatty acids and glycerol
        (1) Fatty acids are converted to acetyl-CoA and used in the Krebs cycle
        (2) Glycerol is converted to glyceraldehyde phosphate and used in glycolysis

III. Photosynthesis

A. General information

1. *Photosynthesis* is a process by which light, water, and carbon dioxide are converted to carbohydrates and oxygen
2. Light is a form of electromagnetic energy that can be seen by the human eye
    a. The different colors of light result from different wavelengths; longer wavelengths appear red and shorter wavelengths appear violet
    b. Shorter wavelengths possess more energy than longer wavelengths; for example, violet light has twice the energy of red light
3. Pigments are molecules that absorb a specific wavelength of light; they are often capable of converting light to chemical energy in the form of a chemical bond
4. The most important pigment in photosynthesis is chlorophyll
    a. *Chlorophyll a* is found in all photosynthetic organisms (plants and cyanobacteria); it must be present for photosynthesis to occur
    b. *Chlorophyll b* is found only in vascular plants, bryophytes, and certain algae; it absorbs light and transfers the energy to chlorophyll a
    c. *Chlorophyll c* is found only in species of brown algae and diatoms
5. *Accessory pigments* also play a role in photosynthesis
    a. They capture light energy and pass it to chlorophyll a
    b. Accessory pigments allow the plant to use a larger portion of the light spectrum for photosynthesis than would be available if only chlorophyll were used
    c. Common accessory pigments are the carotenoids (which appear red, orange, and yellow) of many plants and the phycobilins (which appear red) of cyanobacteria and red algae
6. The summary equation for photosynthesis that occurs in plants, algae, and cyanobacteria is: $6\ CO_2 + 6\ H_2O + sunlight \rightarrow C_6H_{12}O_6 + 6\ O_2$

B. Structures

1. Most photosynthesis occurs within the palisade and spongy mesophyll layers of the leaf (see *Cross Section of a Leaf*, page 24, for an illustration)
2. The cells of the mesophyll layer contain numerous chloroplasts
3. Chloroplasts are the organelles of plant cells where photosynthesis takes place
4. The complex internal structure of the chloroplast consists of thylakoids, grana, and stroma
    a. *Thylakoids* are a series of flattened membranes that form sacs within the chloroplast; the thylakoid membranes contain photosynthetic pigments
    b. *Grana* are collections of thylakoids stacked together within the chloroplast

c. *Stroma* is the semiliquid matrix that fills the chloroplast and surrounds the grana

5. **Photosystems** are collections of 250 to 400 pigment molecules embedded within the thylakoid membranes
   a. Each photosystem consists of a series of antenna pigments that collect light energy and pass it to a central reaction center (chlorophyll a), where the energy is used in photochemical reactions
   b. Plants contain two types of photosystems
      (1) Photosystem I has a reaction center with an absorption peak of 700 nanometers (nm) and is known as P_{700}
      (2) Photosystem II has a reaction center with an absorption peak of 680 nm and is known as P_{680}

C. Chemistry

1. Photosynthesis is divided into light-dependent (light) reactions and light-independent (dark) reactions (see *Reactions of Photosynthesis*)
2. *Light reactions* occur within the thylakoids of the chloroplast
   a. The products of light reactions include ATP, NADPH, and oxygen
   b. In photosystem I, light energy absorbed by P_{700} causes the transfer of electrons to a series of electron carriers, which ends in the reduction of the coenzyme $NADP^+$ to NADPH
   c. In photosystem II, light energy absorbed by P_{680} causes the transfer of electrons to another series of electron acceptor molecules (plastoquinone, cytochrome f complex, and plastocyanin), which deliver the electrons to the P_{700} of photosystem I
   d. The electron chain located between the two photosystems is arranged asymmetrically across the thylakoid membranes so that a proton gradient is generated during electron transfer and ATP can be formed from ADP by phosphorylation
   e. Photosystem II accepts electrons derived from water molecules in a process called **photolysis**
      (1) In photolysis, the water molecule is split to yield protons, electrons, and molecular oxygen: $H_2O \rightarrow 2\ H^+ + 2$ electrons $+ 1/2\ O_2$
      (2) Oxygen is produced as a waste product during photolysis and diffuses out of cells as a gas
      (3) Protons (H^+) are used in ATP production or chemiosmosis
      (4) Chemiosmosis occurs as electrons pass through the electron transport chain; protons follow, and energy is harvested
3. *Dark reactions* (also called the **Calvin-Benson cycle**) occur within the stroma of the chloroplast
   a. The products of dark reactions may be exported from the chloroplast to be used in the cytoplasm of the cell
   b. Dark reactions use the products of light reactions (ATP and NADPH) to convert carbon dioxide into glucose and water
   c. The starting and ending compound of the Calvin-Benson cycle is ribulose bisphosphate (RuBP), a 5-carbon sugar with two phosphate groups attached
   d. The chemical reactions of the Calvin-Benson cycle begin with the addition of carbon dioxide to RuBP by the enzyme ribulose-bisphosphate carboxylase (RuBP carboxylase)

Reactions of Photosynthesis

Light reactions (top diagram) produce adenosine triphosphate (ATP), NADPH (a reduced form of the coenzyme nicotinamide-adenine dinucleotide phosphate), and oxygen. Dark reactions (bottom diagram), also called the Calvin-Benson cycle, use the ATP and NADPH produced by light reactions to convert carbon dioxide into glucose.

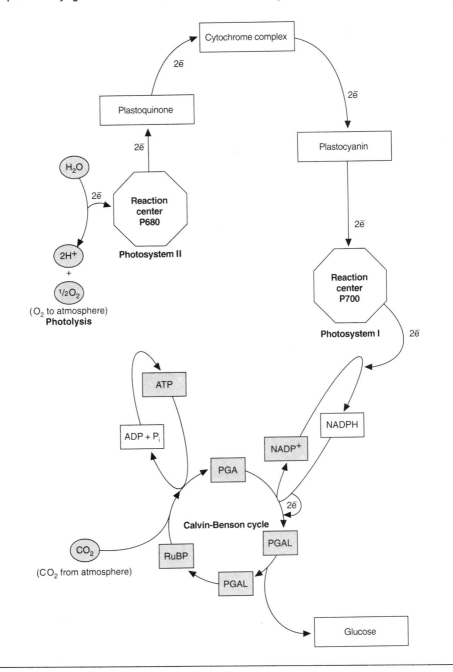

e. The resultant molecule splits into two molecules of phosphoglycerate (PGA): $CO_2 + RuBP \rightarrow 2\ PGA$

f. As PGA molecules pass through the remainder of the cycle, they are chemically altered to form glyceraldehyde phosphate (PGAL), which is subsequently combined to form glucose (the end product of photosynthesis) and to regenerate RuBP

g. Glucose can be converted to starch in the chloroplast for storage

D. Photorespiration

1. **Photorespiration** is a series of chemical reactions that consume oxygen and release carbon dioxide in the presence of light

   a. Because it is not accompanied by oxidative phosphorylation, photorespiration yields no ATP

   b. Because it destroys glucose formed during photosynthesis, photorespiration has a negative impact on the energy efficiency of plants

2. The overall chemical reaction of photorespiration is: $2\ RuBP + O_2 \rightarrow$ phosphoglycolic acid $\rightarrow CO_2 + 3\ PGA$; for every 10 carbon atoms (two 5-carbon molecules in the form of RuBP), three PGA molecules are retained (accounting for 9 carbon atoms) and one is lost as CO_2

3. Photorespiration only occurs in 3-carbon (C_3) plants, so named because the first product of the light reaction is PGA (a 3-carbon molecule)

   a. Normally, about 50% of the carbon fixed (changed into glucose) in C_3 plants by photosynthesis is reoxidized to carbon dioxide in photorespiration, effectively robbing the plant of energy harvested during photosynthesis

   b. Photorespiration occurs when oxygen levels are high and carbon dioxide levels are low in the palisade mesophyll cells; at this point, the enzyme RuBP carboxylase catalyzes the addition of oxygen to RuBP to begin photorespiration instead of catalyzing the addition of carbon dioxide to RuBP to begin the Calvin-Benson cycle

   c. High oxygen and low CO_2 levels occur when plants are under water stress and their stomata are nearly closed; these conditions can result in significant photorespiration

E. Alternative metabolic pathways

1. The two most important photosynthetic adaptations to counter photorespiration are the Hatch-Slack metabolic pathway (4-carbon [C_4] photosynthesis) and **Crassulacean acid metabolism** (CAM)

2. The *Hatch-Slack pathway* eliminates photorespiration through the physical separation of the carbon dioxide storage mechanisms of the C_4 pathway

   a. Hatch-Slack pathway plants are C_4 plants, so named because the first product of carbon fixation is oxaloacetic acid (a 4-carbon molecule)

      (1) C_4 plants are found primarily in the tropics, although some are found in temperate areas; examples include corn, sorghum, sugarcane, and crabgrass

      (2) The leaves of C_4 plants have a distinctive tissue arrangement known as Kranz anatomy

         (a) *Kranz anatomy* is a wreath-like arrangement of mesophyll cells around a layer of bundle sheath cells

         (b) Together these cells form concentric layers around the leaf veins

   b. The Hatch-Slack metabolic pathway has several steps

(1) In the mesophyll cells, CO_2 is added to phosphoenolpyruvate (PEP) by the enzyme PEP carboxylase to form oxaloacetic acid (OAA)

(2) OAA is rapidly reduced within the mesophyll layer to form malate or asparate

(3) Malate or asparate is transported from the mesophyll cells to the bundle sheath cells, where it is decarboxylated (chemical removal of the -COOH group) to form carbon dioxide and pyruvic acid

(4) The CO_2 enters the Calvin-Benson cycle, and the pyruvic acid returns to the mesophyll to regenerate PEP

3. *Crassulacean acid metabolism* decreases the impact of photorespiration by separating CO_2 storage from the fixation of CO_2 in the Calvin-Benson cycle

a. CAM plants are generally desert succulents, such as cacti and stonecrops, but also include some nonsucculents, such as pineapples, orchids, spanish moss, quillwort, and some ferns)

b. CAM plants have photosynthetic cells within their leaves that can store CO_2 at night, when the enzyme PEP carboxylase combines PEP and CO_2 to form malic acid

(1) At night, the malic acid is accumulated and stored in a vacuole within the cell

(2) During the day, the malic acid is decarboxylated to release CO_2 and PEP, supplying CO_2 to the Calvin-Benson cycle within the same cell in which it was stored at night

c. The stomata of CAM plants open at night to accumulate CO_2 when temperatures are cooler and less water is lost from transpiration; they remain closed during the day to prevent excessive water loss

F. Photosynthetic products

1. The glucose made during photosynthesis supplies energy for all plant activities; it also supplies the carbon fragments needed to synthesize all the major cellular organic molecules (polysaccharides, lipids, proteins, and nucleic acids)

2. Approximately half of the glucose produced by photosynthesis is used as an energy source for cellular respiration in the mitochondria of nonphotosynthetic tissues (including roots, flowers, and fruits) and in all cells at night, when photosynthesis cannot occur

3. Glucose formed during photosynthesis is transported from the leaf to all other plant organs as the disaccharide sucrose

Study Activities

1. Discuss the principal coenzymes involved in cellular respiration and photosynthesis.

2. Diagram the process of cellular respiration, showing the materials needed and the resulting end products.

3. Compare and contrast aerobic and anaerobic respiration.

4. Diagram the process of photosynthesis, showing both light and dark reactions.

5. Identify the structures of C_3 and C_4 leaves, and discuss their relationship to photosynthetic reactions.

5

Plant Hormones and Their Actions

Objectives

After studying this chapter, the reader should be able to:
- List the principal plant hormones and explain their effects.
- Discuss tropisms in plants.
- Explain the role of auxin in phototropism and geotropism.
- Discuss how the photoperiod affects flowering in plants.
- Describe the role of phytochrome.

I. Plant Hormones

A. General information
1. A *hormone* is a substance produced in plant tissue that has a specific effect when transported to another area of the plant
2. Most plant hormones are stimulatory (promote some effect), but they can also be inhibitory or become inhibitory at high concentrations
3. Plant hormones often work collectively, and the effects of one hormone are usually influenced by the presence of other hormones
4. Plant hormones are produced in small quantities and are involved in the regulation of plant growth and development
5. The external environment can cause a change in the concentration, distribution, or sensitivity of a hormone and thus alter the hormone's effect
6. The effect of a plant hormone depends on the particular tissue or organ on which it is acting
7. The major plant hormones are auxin, cytokinins, gibberellins, ethylene, and abscisic acid

B. Auxin
1. Depending on the particular plant tissue, auxin can promote growth by increasing cell elongation (in most tissues), promoting cell division in the vascular cambium, stimulating cell differentiation in vascular stem tissue, promoting the growth of fruits, and preventing the loss of leaves or fruit
    a. Cell elongation occurs because auxin increases the malleability of the cell wall and allows ***turgor pressure*** to increase the size of the cell
    b. The increased plasticity appears to result from auxin's ability to promote the breaking of bonds within the cell wall

2. *Auxin* is a collection of hormones, which include naturally occurring indoleacetic acid and synthetic phenylacetic acid and 4-chloro-indoleacetic acid

3. It is produced in meristems and slowly transported (at a rate of 0.5 to 1.5 cm/hour) from cell to cell in phloem parenchyma and the parenchyma surrounding the vascular bundles

4. Auxin's effects depend on its concentration; moderate amounts are generally stimulatory and large amounts may be inhibitory

5. Different plant parts respond differently to auxin concentrations; for example, roots are generally more sensitive than stems to lower concentrations of auxin

6. Auxin is involved in a number of plant responses

   a. Cell division in the vascular cambium increases under the influence of auxin, causing the cambium to produce xylem and phloem

   b. *Apical dominance* occurs when auxin (and cytokinins) from the apical meristem inhibit the growth of lateral buds; the result is a growth pattern in which the lateral branches near the top of the plant are shorter than those near the ground, giving the plant an overall conical or pyramidal shape (such as pine trees)

   c. The ovary wall develops into the fruit under the influence of auxin from developing seeds

   d. Phototropism and geotropism are caused by the unequal distribution of auxin throughout the plant

C. Cytokinins

1. *Cytokinins* promote mitosis, cell enlargement, differentiation of tissues, development and retention of chlorophyll in leaves, stimulation of cotyledon growth in seeds, and slow aging in leaves

2. These hormones are found in actively dividing tissues of seeds, fruits, and roots

   a. They are particularly abundant in roots, young leaves, and developing fruits

   b. They are found in high concentrations in the roots of young seedlings

3. Cytokinins are thought to be transported in xylem tissue

4. Approximately 18 different naturally occurring cytokinins have been isolated in plants; kinetin was the first cytokinin to be discovered

5. Cytokinins have several effects

   a. They influence protein synthesis through involvement in the attachment of transfer ribonucleic acid (tRNA) to messenger RNA (mRNA) and the ribosome

   b. They have a stimulatory effect on cell division, which appears to be greatly influenced by the presence of and interaction with other hormones, such as auxin, and the influence of external factors, such as the presence of light

   c. Along with auxins, they influence apical dominance

   d. They also influence the establishment of areas within the plant that preferentially attract and concentrate nutrients

D. Gibberellins

1. *Gibberellins* promote the extension of stems, break dormancy in seeds, and promote growth in embryos and seedlings

2. They are present in varying amounts in all plant parts, although the highest concentrations are found in immature seeds

3. Gibberellins are transported in xylem and phloem tissues

4. Fifty-two types of gibberellins have been isolated in plants

5. The biologic activity of gibberellins is often compared with that of auxins; both hormones promote cell elongation and cambial activity and stimulate nucleic acid synthesis and protein synthesis, although the mode of action is likely to be different
6. Gibberellins also can cause premature and rapid growth of stems in some plants (*bolting*)

E. Ethylene

1. **Ethylene** promotes the maturation and ripening of fruits, influences the release of seeds, speeds the aging of floral parts, induces flowering, and inhibits leaf growth (by promoting leaf senescence and abscission)
2. It is the only plant hormone that is a gas; its chemical formula is $CH_2{=}CH_2$
3. Ethylene is produced in many healthy plant tissues; it is diffused rapidly
4. Ethylene production is stimulated by wounding, rubbing, radiation, and the presence of auxin
5. Many of the effects originally attributed to auxin alone appear to involve the interaction of auxin and ethylene
6. In most fruits, the rate of respiration undergoes a sharp rise and fall near the end of ripening, a phenomenon referred to as **climacteric**
   a. Climacteric triggers changes that ripen a fruit
   b. The amount of ethylene present in fruit tissue increases a hundredfold just before and again during climacteric
   c. Conditions that slow or prevent ripening, such as low temperatures, retard ethylene production
   d. The application of ethylene to unripened fruit can bring on climacteric and accelerate ripening
   e. The actual mechanism by which ethylene promotes fruit ripening is unclear; current theories suggest that it alters cell membrane permeability, stimulates enzyme synthesis, or both
7. In dicot seedlings, ethylene is produced in the plumule or hypocotyl regions and appears to stimulate extension of the hypocotyl so that the seedling emerges from the soil

F. Abscisic acid

1. **Abscisic acid** inhibits seed germination and protein synthesis and accelerates abscission; it differs from other plant hormones because it is usually inhibitory, not stimulatory
2. Abscisic acid is a unique compound composed of a single carbon ring with an associated carbon-hydrogen chain
3. It is produced primarily in mature leaves and the root cap and is often found in dormant tissues
4. Abscisic acid appears to be transported in xylem and phloem tissues
5. Its most important role is its control of stomatal closing
   a. It appears to inhibit potassium uptake by **guard cells**, which results in a loss of guard cell turgor and closure of the stomata
   b. It accumulates in plants during times of water-related stress and drought conditions, resulting in stomatal closure to conserve water
6. Abscisic acid also inhibits seed development and germination by suppressing the synthesis of enzymes necessary for germination

7. The role of abscisic acid in promoting dormancy in buds is not well established and remains controversial
8. Abscisic acid is considered by some botanists to be involved (along with auxin) in the response of roots to gravity (geotropism)
9. The mechanism of action of abscisic acid is not well understood

G. Applications
1. Plant hormones have many practical applications
2. Ethylene and auxin in particular have numerous commercial uses
    a. Bananas and lemons become ripe during transport after exposure to ethylene
    b. Flowering in plants, such as pineapples and ornamental bromeliads, is stimulated by applying ethylene
    c. White potatoes are often sprayed with auxin to inhibit sprouting of buds, which increases the storage time
    d. Apple, pear, and citrus trees are often sprayed with auxin to prevent the fruits from dropping; auxin treatment delays fruit drop for several days and allows for more efficient harvesting
    e. Auxin in the form of "root powders" is applied to cuttings to stimulate root growth
    f. Some selective herbicides, such as those for broad-leaf weeds, are synthetic auxins; auxin herbicides overstimulate plants, causing uncontrolled growth and eventual plant starvation

II. Plant Movements

A. General information
1. Plant movements commonly result from growth responses (*tropisms*) or changes in internal osmotic pressures (turgor movements)
2. Plant hormones typically are involved in initiating or controlling plant movements, particularly tropisms

B. Tropisms
1. *Tropisms* are growth responses that result in the curvature of the plant or plant organs (stems, leaves, or flowers) toward or away from an external stimulus
    a. Positive tropisms occur if growth is toward the stimulus
    b. Negative tropisms occur if growth is away from the stimulus
2. They are caused by differences in the rate of cell elongation on opposite sides of the plant organ; auxin or acetic acid promote these differences in cell elongation
3. The three types of tropisms displayed by plants are phototropism, geotropism, and thigmotropism
    a. *Phototropism* is the growth of a plant shoot toward light (see *Phototropism*, page 48)
        (1) Shoot cells located on the side not receiving any light elongate faster than those on the illuminated side because of the asymmetric distribution of auxin, which moves away from the light toward the dark side of the stem

Phototropism

Auxin is produced in the apical meristem and transported down the stem through vascular tissue. When the stem is illuminated from the side, auxin (shaded area) accumulates on the darker side and stimulates cell elongation, which bends the shoot toward the light source.

Transport of auxin

Auxin moves from illuminated side

Auxin promotes cell elongation on dark side

Shoot bends toward light source

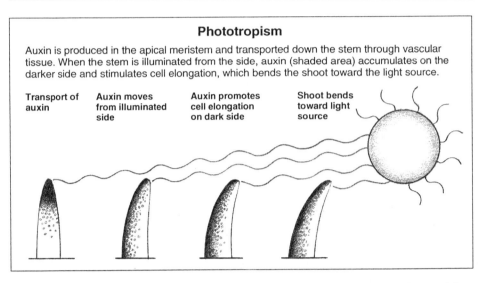

(2) The unequal growth causes the plant or plant organ to bend toward the light

(3) Light wavelengths from 400 to 500 nanometers (nm) are most effective in inducing phototropism

b. **Geotropism**, or **gravitropism**, is the growth response of roots or shoots to gravity (see *Geotropism in Roots*)

(1) Roots display positive geotropism (growth toward the force of gravity), and shoots display negative geotropism (growth away from the force of gravity)

(2) In certain cells, the amyloplasts (plastids containing starch) serve as gravity-sensing bodies known as **statoliths**

(a) In roots, statoliths are found only in the central core of the root cap; in stems, they are found in cells adjacent to the vascular bundles

(b) Statoliths respond to gravity by settling against the cell membrane that is facing downward

(c) Collections of statoliths cause hormones (such as auxin) to accumulate in the tissues, which in turn promotes cell elongation and the bending of the plant or plant organ

(d) The mechanism by which statoliths stimulate cell elongation is not yet understood

(3) Stems bend upward because auxin and gibberellin stimulate cells on the lower side of the stem to elongate faster than those on the top

(4) Roots curve downward because auxin and gibberelins accumulate on the lower side of the root, which inhibits cell elongation

c. **Thigmotropism** is directional growth in response to touch or physical contact with other structures

(1) Thigmotropism occurs because cells touching an external structure shorten while those on the other side elongate; the response can occur rapidly (in less than one hour)

(2) It is usually restricted to tendrils (coiled, stringlike, clasping or anchoring structures) found on the stems of vines, such as ivy

Geotropism in Roots

Growing root tips detect gravity by the position of starch-containing plastids known as statoliths. When statoliths accumulate at the bottom of a cell, normal upright growth takes place. When statoliths accumulate on the side of a cell, asymmetrical growth and subsequent curvature of the root tip occurs.

Root tip in vertical position grows normally

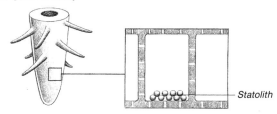

Statolith

Root tip in horizontal position curves

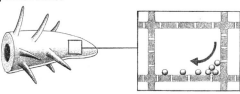

(3) Auxin may play a role in this tropism, but the exact mechanism of action is not known

C. Turgor movements

1. *Turgor movements* are reversible movements caused by changes in turgor pressure within specialized plant cells
2. Turgor pressure results from the uptake of water by the cells
   a. Changes in turgor pressure within cells are generally restricted to specialized structures called *pulvini*
   b. Pulvini are jointlike swellings located adjacent to the movable plant organ; they are generally found only in leaves or floral parts
   c. When pulvini cells experience changes in turgor, the plant part moves (that is, the leaf changes position or the flower is reoriented)
3. Examples of turgor movement are leaf movements, sleep movements, and solar tracking
   a. *Leaf movements*, which occur in the mimosa (sometimes nicknamed the "sensitive plant") and such carnivorous plants as the Venus flytrap, are characterized by rapid movement of the leaves, generally taking only a few seconds
   b. *Sleep movements*, which occur in the prayer plant and the common wood sorrel, are the daily movement of leaves from a vertical position in the evening and to a horizontal position in the morning
   c. *Solar tracking*, which occurs in clover, soybeans, sunflowers, and cotton, is the movement of leaves toward the sun during the course of each day

III. Photoperiodism

A. General information

1. Photoperiodism is a physiologic response to day length
2. It governs the seasonal aspects of a plant's life cycle, such as flowering, fruit production, seed formation, and dormancy in woody plants
3. Phytochrome allows plants to measure day length

B. Phytochrome

1. **Phytochrome**, a pigment composed of proteins, measures the relative amounts of light and dark to which a plant is exposed
2. Phytochrome occurs in two forms, designated P_r (red absorbing, 660 nm) and P_{fr} (far-red absorbing, 730 nm)
   a. Plants synthesize P_r, which remains P_r if plant is kept in the dark
   b. During the day, when phytochrome is exposed to red light, P_r is converted to P_{fr}
   c. At night, P_{fr} gradually and spontaneously reverts to P_r or is converted to P_r when exposed to far-red light (700 to 800 nm)
   d. Plants do not use the disappearance of P_{fr} to directly measure night length
      (1) The conversion of P_{fr} to P_r is completed within a few hours after sunset and is affected by temperature
      (2) Plants therefore use the ratio of P_r to P_{fr} to trigger biologic activity, such as flowering
      (3) The P_r:P_{fr} ratio synchronizes the plant's internal biologic clock, which governs the life cycle of the plant; however, the biologic clock and its mechanism are not well understood

C. Flowering

1. Early studies indicated that flowering in plants is a response to the **photoperiod** (length of daylight), which prompted botanists to classify plants as short-day, long-day, or day-neutral
   a. **Short-day plants** flower in late summer, fall, and winter, when the days are short; examples include asters, chrysanthemums, poinsettias, dahlias, goldenrods, ragweeds, strawberries, and violets
   b. **Long-day plants** flower in late spring and summer, when the days are long; examples include beets, clover, corn, gladiolus, wheat, barley, potatoes, spinach, and lettuce
   c. **Day-neutral plants** are not affected by the photoperiod; examples include beans, carnations, cyclamens, cotton, roses, snapdragons, sunflowers, tomatoes, and dandelions
2. Subsequent studies found that the length of uninterrupted darkness (not the length of light) was the critical factor in flowering
   a. Short-day plants exposed to 16 hours of darkness and 8 hours of light in a 24-hour cycle flower normally; if the period of darkness is interrupted briefly by a flash of light, these plants do not flower, even though they receive the critical amount of light
   b. Long-day plants exposed to 16 hours of light and 8 hours of darkness in a 24-hour cycle flower normally; they also flower if the period of darkness is interrupted by a brief exposure to light, as long as they receive the critical amount of light

3. *Florigens* are a family of hormones that stimulate flowering
   a. Florigens are produced in leaves when the plant is exposed to the appropriate photoperiod (which is determined by whether the plant is a short-day or long-day type)
   b. They are then released from the leaves and travel to the apical meristems, where they convert the apical bud from stem growth to flower formation

D. Other physiologic responses
   1. Seed germination is affected by phytochrome activity
      a. Germination in some seeds is inhibited by exposure to light to ensure that the seed germinates only if covered by soil
      b. Germination in other seeds is stimulated by light to ensure that the seed is close enough to the soil surface to successfully germinate and emerge before energy reserves are exhausted
   2. Seedling growth depends on phytochrome
      a. During growth immediately after germination, the seedling beneath the soil surface appears etiolated (yellow to colorless, small leaves, and spindly stem growth) because of the lack of light
      b. After the seedling emerges from the soil, the detection of light by phytochrome activity results in chlorophyll production and initial leaf growth
   3. Growth of shaded plants also depends on phytochrome
      a. Plants growing in the shade of other plants receive light rich in the 700- to 800-nm range
      b. Light in this range causes a dramatic upward shift in the P_r:P_{fr} ratio, which stimulates stem elongation

Study Activities

1. Name the five principal plant hormones and describe their effects.
2. Explain the role of the statoliths in geotropism.
3. Describe the conditions under which short-day, long-day, and day-neutral plants flower.
4. Explain how cell elongation occurs in shaded plants.

6

Soils and Plant Nutrition

Objectives

After studying this chapter, the reader should be able to:
• Understand the importance of organic matter in soils.
• Describe the processes by which soils are formed.
• List the principal soil horizons and describe their features.
• Describe the macronutrients and micronutrients needed by most plants.
• Recognize and diagnose common plant nutrient deficiencies.

I. Soils

A. General information
    1. Soil is a complex mixture of tiny particles of inorganic and organic matter, water, air, and living organisms
    2. It anchors plants and provides a reservoir for the nutrients and water needed by plants for growth and development
    3. About 50% of the total volume of soil consists of space between individual soil particles; this space is occupied by variable amounts of water and air
        a. In extremely dry soils, the space is occupied by air
        b. In waterlogged soils, the space is occupied by water

B. Components
    1. *Inorganic matter*, which is not derived from living things, is classified according to its importance to plants
        a. *Macronutrients* are required by plants in large amounts; examples include calcium, carbon, hydrogen, magnesium, nitrogen, oxygen, phosphorus, potassium, and sulfur
        b. *Micronutrients*, or trace elements, are required by plants in small amounts; examples include boron, chlorine, copper, iron, manganese, molybdenum, and zinc
    2. *Organic matter*, which is derived from living things, consists of dead leaves, stems, and roots; insect remains; animal droppings; and worm secretions accumulated in the upper portion of the soils
        a. Soils contain 1% to 7% organic matter by weight
        b. Besides providing nutrients, organic matter retains moisture

c. Bacteria, fungi, worms, and other small organisms break down (by digestion) organic compounds into simpler forms, such as nitrate, phosphate, potassium, and sulfate—all of which can be used by plants

d. Partially decomposed organic matter is called **humus**

3. Water is held in the spaces between soil particles or adheres to soil particles

a. The ability of a soil to hold water against the force of gravity depends on porosity (number of pores or open spaces in the soil) and soil particle size

(1) A porous soil (one with many open spaces between large particles) cannot hold water because it drains through quickly

(2) A soil with many small particles holds more water because the soil's surface area increases as particle size decreases

b. Water-holding capacity cannot be equated with water availability; some water molecules may adhere so strongly to soil particles that they are not available to plants

c. Two common methods used to measure soil water-holding capacity are field capacity and wilting coefficient

(1) *Field capacity*, the percentage of a soil's weight that is water, is a measurement of the amount of water soil particles can hold

(2) *Wilting coefficient*, or wilting point, is the percentage of water remaining in the soil when a plant is permanently wilted and cannot recover

(3) In soils with predominately smaller particles, the total surface area of the soil is relatively large, and both the field capacity and wilting coefficient are high

(4) In soils with predominately larger particles, the total surface area of the soil is relatively small, and both the field capacity and wilting coefficient are low

(5) The water available to plants is equal to the field capacity minus the wilting coefficient

4. *Air* is also found in the spaces between soil particles

a. It is a mixture of gases similar to atmospheric air, but it may contain a higher concentration of carbon dioxide

b. Carbon dioxide produced by the respiration of plant roots and soil organisms can reach concentrations of 10% in the soil, compared to 0.03% in the atmosphere

5. *Living organisms* are important components of soils, even though they make up only about 0.1% of the soil mass

a. The most common soil organisms are bacteria, fungi, molds, protozoa, mites, nematodes, earthworms, insects, and burrowing animals

b. Living organisms churn and mix the soil and facilitate the breakdown of organic materials

C. Development

1. The ultimate source of soils are the rocks and minerals of the earth

2. These parent materials must be broken down to form soils

a. *Physical weathering*, or mechanical breakdown, breaks the parent rock into smaller pieces by exposure to temperature changes and the physical action of moving ice and water, growing roots, and human activities

b. *Chemical weathering* is the dissolution of the parent rock by exposure to rain, surface water, and such gases as oxygen

3. Plants play a significant role in soil development

a. A plant's growing roots can exert enough pressure to split rocks and begin the weathering process

b. Certain plants, such as lichens and mosses, secrete acids that dissolve rocks and penetrate and loosen rock particles

c. Once anchored and growing, plants can trap windblown materials (such as leaves) and thereby augment soil development

d. As they die and decompose, plants add organic matter to developing soil

D. Properties

1. The *physical properties* of soil consist of texture and structure

  a. *Texture* is determined by the size and relative proportion of soil particles

   (1) Soils contain a variety of particle sizes; clay particles are less than 0.002 millimeters (mm) in diameter, silt particles are between 0.002 and 0.02 mm in diameter, and sand particles range from 0.02 to 2 mm in diameter

   (2) Soils are characterized by the predominate particle type; for example, clay soils contain mostly clay particles

   (3) Soils best suited for agricultural applications (known as loams) are those with approximately equal proportions of clay, silt, and sand

  b. *Structure* is the way that soil particles clump together

   (1) Soil structure largely depends on the amount of clay and organic material in the soil

   (2) Clay soils hold together strongly and, when wet, can form massive clumps (this is why potters' clay holds together and can be molded)

   (3) A soil with too much clay impedes root growth; one with too little clay cannot hold plants upright

2. The *chemical properties* of soils are determined primarily by the interaction of pH, water, and available plant nutrients

  a. pH is a measure of the acidity or alkalinity of a substance; it is formally defined as the negative logarithm of the hydrogen ion concentration

   (1) The pH scale ranges from 0 to 14; a pH of 7 is considered neutral, while a pH between 0 and 6 is acidic and one between 8 and 14 is alkaline, or basic

   (2) Acidic soils (soils with a low pH) contain large numbers of hydrogen ions, which are formed when carbon dioxide in the atmosphere reacts with water to produce carbonic acid

   (3) These hydrogen ions can displace soil nutrients, such as calcium, potassium, and sodium, from soil particles, thereby increasing the possibility of *leaching* (washing away of dislodged nutrients as water moves through the soil)

   (4) In alkaline soils (soils with a high pH), some ions, such as iron, may precipitate out of solution and therefore be unavailable to plants

   (5) The optimal balance of nutrients is found in soil with a neutral pH

  b. Positively charged ions (cations) attach to clay particles in soil in a process called *cation exchange*

   (1) Cation exchange is necessary for plants to obtain nutrients; however, leaching may occur once the ions are released from the soil particles

   (2) The chief cations important in plant nutrition are calcium, magnesium, potassium, and sodium

(3) Plant roots can displace cations by producing and releasing hydrogen ions; roots may also exchange hydrogen ions for cations

(4) The cation exchange capacity of a soil increases as the percentage of clay particles increases

c. Negatively charged ions (anions) respond differently because they do not attach to clay particles in soil

(1) The chief anions important in plant nutrition are nitrate (NO_3^-), sulfate (SO_4^{-2}), carbonate (HCO_3^-), phosphate (HPO_4^{-2} or $H_2PO_4^-$), and hydroxyl ions (OH^-)

(2) Anions leach more rapidly than cations since they do not attach to clay particles; the exception is phosphate, which forms an insoluble precipitate and is absorbed or combined with compounds containing iron, aluminum, and calcium

d. In areas of low rainfall, such as the semiarid valleys in the western and southwestern regions of the United States, calcium and other alkaline compounds are not leached away; as a result, soils in these areas may become too alkaline (pH of 7.5) for plant growth

E. Profile

1. The soil profile is a series of distinctive horizontal layers called *soil horizons*

2. Most soils contain at least three or four of the six soil horizons (see *Soil Profile*, page 56, for an illustration)

a. The *O horizon* is the uppermost layer; it contains surface litter (freshly fallen leaves, organic debris, and partially decomposed organic matter)

b. The *A horizon* is the topsoil and contains decomposed organic matter, plant roots, living organisms, and some inorganic minerals

c. The *E horizon* is the zone of leaching, where dissolved or suspended materials are carried downward by water

d. The *B horizon* is the subsoil, in which iron, aluminum, organic compounds, and clay accumulate from the upper layers; the deposition of these substances produces distinct colors within this horizon

e. The *C horizon* is the parent material that forms the upper layers through weathering and consists of partially broken down inorganic materials

f. The *R horizon* is the bedrock, a layer of rock that is mostly impenetrable except for fractures

II. Plant Nutrition

A. General information

1. Plants require a variety of minerals to survive, even though they harvest energy from sunlight through photosynthesis

2. Essential elements are those required by plants to complete their life cycle

a. Elements in ionized form or in solution can be used by a plant because they are readily available to the plant

b. Elements that adhere to soil particles or become precipitates are unlikely to be available to a plant

3. Common manifestations of nutrient deficiency are a lack of chlorophyll (***chlorosis***) in leaves or stems, necrosis (dead patches of tissue), stunted growth, wilting, and deformities

Soil Profile

The depth of the O and A horizons varies with vegetation type and climate. Most living organisms (animals, bacteria, and plant roots) within a soil are located in the O and A horizons.

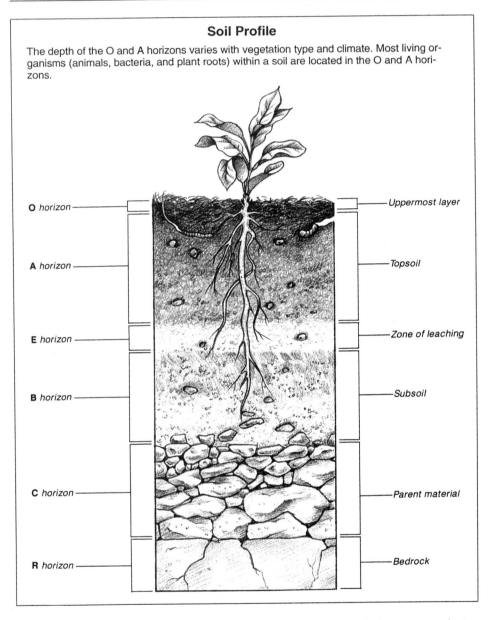

O horizon — — Uppermost layer

A horizon — — Topsoil

E horizon — — Zone of leaching

B horizon — — Subsoil

C horizon — — Parent material

R horizon — — Bedrock

4. Besides the macronutrients and micronutrients needed by all plants, some plants also require sodium, aluminum, silicon, chlorine, gallium, and cobalt

B. Macronutrients
1. Nitrogen, phosphorus, sulfur, potassium, calcium, and magnesium are macronutrients, those that are required by plants in large amounts
2. Nitrogen is an essential part of plant proteins, amino acids, and nucleic acids
   a. Plants obtain nitrogen from soils in the form of nitrate ions (NO_3^-) and, to a lesser degree, ammonium ions ($NH4^+$)

b. Nitrogen deficiency usually results in yellow leaves and stunted growth

c. Excess nitrogen results in vigorous vegetative growth (numerous leaves and large stems) and suppression of storage organs, fruits, and seeds

3. Phosphorus is a component of nucleic acids, cell membranes, and adenosine triphosphate

a. It is important in the production of fruits and seeds and the growth of root and shoot tips

b. Plants often obtain phosphorus from potash fertilizers

c. Phosphorus deficiency results in stunted growth and leaves with purplish veins

4. Sulfur is a component of proteins and amino acids

a. Plants obtain sulfur as sulfate ions

b. Sulfur deficiency results in yellowing and dead spots on leaves and decreased root growth

5. Potassium serves as an enzyme activator and plays a role in stomatal control

a. It is concentrated in areas of active growth

b. Potassium deficiency results in yellowing and browning of leaves

6. Calcium is important in cellular and tissue processes

a. It is an element of the middle lamella of the cell wall and contributes to cell membrane integrity

b. Calcium is needed for assembly of the spindle fibers during cell division; without calcium, cells cannot divide

c. Calcium deficiency results in the death of growing tissues

7. Magnesium is the chief component of chlorophyll

a. It also serves as an enzyme cofactor or activator

b. Magnesium deficiency results in the formation of dead spots on young leaves

C. Micronutrients

1. Iron, boron, zinc, copper, molybdenum, and manganese are micronutrients, those that are required by plants in small, or trace, amounts

2. Micronutrients generally serve as cofactors and electron carrier molecules

3. Iron is found in cytochromes and coenzymes of respiration and photosynthesis

4. Boron regulates carbohydrate breakdown

5. Zinc, copper, molybdenum, and manganese are important enzyme activators

6. A deficiency of one or more of these micronutrients decreases metabolism and growth and weakens the plant, making it susceptible to disease

Study Activities

1. Define field capacity and wilting coefficient.

2. Discuss how soil is formed.

3. List and describe the principal physical and chemical properties of soils.

4. Name and describe the six soil horizons that make up a soil profile.

5. Prepare a chart that lists the macronutrients and micronutrients needed by plants. Include the importance of each nutrient and the possible consequences of deficiency.

7

Absorption, Transport, and Transpiration

Objectives

After studying this chapter, the reader should be able to:
- Trace the pathway of water in a plant.
- Describe apoplastic and symplastic transport in roots.
- Explain the role of the endodermis in determining what materials enter the xylem.
- Discuss the benefits of transpiration.
- Describe the mechanism of stomatal opening and closing.
- Explain the movement of materials in the processes of transpiration and translocation.

I. Absorption and Transport by Roots

A. General information
1. Absorption of water and dissolved nutrients (minerals and elements from soil) takes place in the root hairs
2. These materials are absorbed by an active transport process and by passive transport processes, such as diffusion and osmosis
   a. *Active transport* is an energy-consuming process in which a solute is transported across a membrane in a direction of increasing concentration, often a direction opposite that of diffusion
      (1) Absorption of ions at the epidermal surface is an active process
      (2) In waterlogged soils, plants wilt due to low oxygen levels, which interferes with the root's ability to conduct respiration and produce the adenosine triphosphate (ATP) necessary for active transport
   b. *Diffusion* is the tendency for molecules of any substance to spread out into the available space; although each molecule moves randomly, the diffusion of a group of molecules may be directional
      (1) In diffusion, molecules move from an area of higher concentration to an area of lower concentration
      (2) Diffusion continues until a difference in concentration no longer exists
      (3) This passive process tends to distribute particles uniformly throughout a medium
   c. *Osmosis* is the movement of water or any solvent across a selectively permeable membrane (one through which only certain materials can pass); in the absence of other forces, water moves from an area of higher water concentration to an area of lower water concentration

B. Absorption

1. Water and dissolved nutrients (minerals and elements from soil) are absorbed through root hairs (projections of the epidermis found near root tips) and epidermal cells
2. Most mineral nutrients are moved into a root by active transport
3. Once the concentration of solutes (dissolved nutrients) in the root cells is increased by active transport of minerals, water may passively diffuse into the root by osmosis
4. Most of the water absorbed from the soil enters through the younger parts of the root

C. Transport pathways

1. When water first enters the root, it follows a specific pathway
   a. Water flows across the cell membranes of root hairs or epidermal cells into the root cortex
   b. It then enters the stele by passing through the endodermis and pericycle
   c. Once in the stele (vascular cylinder), water enters the xylem and moves upward to the stem
2. Water (or solutes) entering the root cortex are transported via the symplastic or apoplastic pathway (see *Apoplastic and Symplastic Transport*, page 60)
   a. The **symplastic pathway** transports materials within the cytoplasm of the root cortex cells via cytoplasmic connections called plasmodesmata
      (1) This pathway enables water and dissolved minerals to be transported from cell to cell
      (2) The ultimate destination is the xylem tissue of the vascular cylinder
   b. The **apoplastic pathway** transports materials within the cell walls and intercellular spaces between the root cortex cells
      (1) It is interrupted at the endodermis by a barrier (Casparian strip) formed by the endodermal cells
      (2) The Casparian strip diverts water and solutes from intercellular spaces into endodermal cells, permitting them to control the movement of materials from the root cortex into the vascular cylinder, thereby determining which materials enter the xylem

D. Root pressure

1. The development of pressure within the roots (**root pressure**) forces water and dissolved nutrients upward through the xylem
2. Root pressure develops from the accumulation of solutes and water entering the stele
   a. When a plant is cut at a junction of the root and shoot, water continues to exude from the cut because of root pressure
   b. If roots are killed or deprived of oxygen, root pressure drops
3. Root pressure can reach between 30 and 45 pounds per square inch and provides a mechanism for filling the xylem vessels
   a. It is important in vines, whose vessels are emptied of water during the winter season
   b. It is also important in herbaceous plants on hot, dry days, when xylem vessels may need to be refilled
4. Root pressure is only a subsidiary mechanism of water movement in most plants
   a. It cannot move the large volumes of water needed in larger plants
   b. It appears to operate in smaller plants when transpiration is low

Apoplastic and Symplastic Transport

In apoplastic transport (A), water passes through the root cortex by traveling in the intercellular spaces between cells. In symplastic transport (B), water passes through the root cortex by moving from cell to cell within the cytoplasm. At the endodermis, water must pass through the endodermal cells because the Casparian strip blocks apoplastic transport.

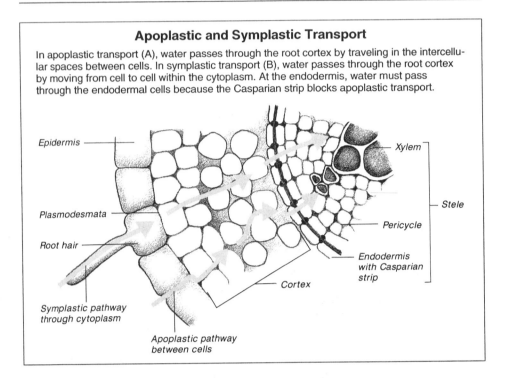

II. Transpiration

A. General information
1. *Transpiration* is the loss of water vapor by plant parts
   a. Water evaporates from the cells lining the intercellular spaces of the plant and diffuses into the atmosphere
   b. Water vapor in the intercellular air spaces is often at 100% relative humidity
2. Although transpiration can occur at nearly any plant surface, most occurs at the stomata and lenticels
   a. *Stomata* are small openings in the epidermis of leaves and stems through which gas exchange occurs; they are the primary sites of transpiration
   b. *Lenticels* are spongy areas on the surface of stems and roots that allow gas exchange between the internal tissues of the plant organ and the atmosphere
   c. *Other plant organs* account for approximately 5% to 10% of total water loss
3. During the course of a growing season, large amounts of water can be lost through transpiration; for example, a tomato plant loses 125 liters (33 gallons) of water and a corn plant loses 206 liters (55 gallons) during a growing season

B. Factors affecting transpiration
1. Transpiration is affected by physical and biologic factors
2. The physical factors include temperature, humidity, wind speed, and light intensity
   a. High temperatures increase transpiration; for every 10 °C (18 °F) increase in temperature, transpiration increases 10% to 20%

     b. High humidity in the surrounding air slows the diffusion of water vapor from the plant to the atmosphere (water vapor moves from an area of high concentration to one of low concentration); low humidity or dry air increases transpiration

     c. Increased wind speed and air currents increases transpiration from plant surfaces

     d. Increased light intensity heats plant surfaces and increases transpiration by raising the rate of evaporation in intercellular spaces

3. The biologic factors include stomatal opening and moisture (water) content

     a. The number and degree of opening of the stomata affects the rate of transpiration; plants with many stomata have a higher rate of transpiration than those with fewer stomata because there are more openings through which water vapor can escape

     b. The water content of plant tissues affects transpiration in two ways

       (1) In plants with low water content, the stomata close to reduce transpiration before the plant becomes dehydrated

       (2) In plants with high water content, the stomata open to encourage transpiration

C. Regulation of transpiration

1. Plants regulate transpiration by opening and closing the stomata

     a. Water from the vascular tissue (veins) of the leaf exits the open stomata

     b. Closing the stomata reduces water loss, but it also prevents the exchange of oxygen and carbon dioxide between the leaf and the atmosphere

2. Stomata provide a mechanism for regulating the conflicting demands of increased carbon dioxide uptake necessary for photosynthesis and the need to conserve water

3. Each stoma consists of an opening in the leaf epidermis bordered on either side by guard cells

     a. Asymmetrical thickening of the guard cell walls and the radial arrangement of cellulose microfibrils within the cell wall causes the cells to bend when they become turgid (filled with water)

     b. Changing turgor pressure (water pressure) within the guard cells opens and closes the stomata

       (1) When guard cells are filled with water, they become turgid and bend, thereby opening the stoma

       (2) When guard cells lose water, they become less turgid and relax to obscure the opening, thereby closing the stoma

4. The mechanism by which turgor within the guard cells is altered is not well understood

     a. An active potassium ion pump may be involved in stomatal operation (the potassium ion concentration is higher in open guard cells than in closed cells)

     b. The potassium pump is run by ATP, which is generated by photosynthesis and cellular respiration within the guard cells

     c. In daylight, potassium ions accumulate in the cytoplasm of the guard cells; as a result, water moves into the cell by osmosis, thereby increasing the cell's turgor and causing the stoma to open

     d. In the dark or when the plant is under water stress (drought), potassium ions leak out (or are pumped out) of the cytoplasm of the guard cells; as a re-

Transpiration Pathway

Water is drawn from the soil through the epidermis or root hairs of the epidermis. It passes into the root cortex, then enters the vascular cylinder and xylem. Xylem vessels carry water from root to shoot and leaves. The water that enters leaf tissues from the xylem eventually exits the leaf via the stomata. The mechanism by which water moves through the plant is called *transpirational pull.*

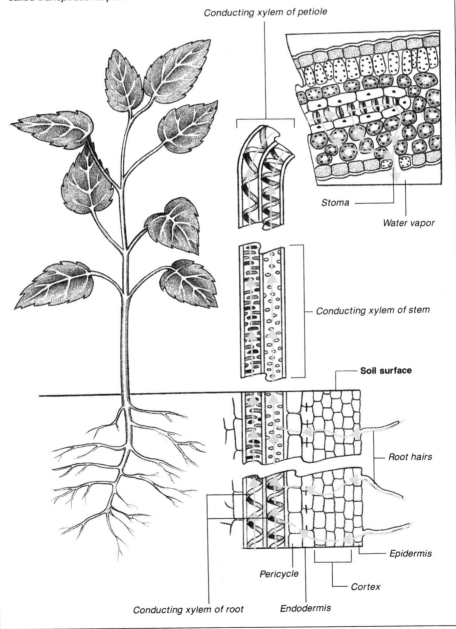

Conducting xylem of petiole

Stoma

Water vapor

Conducting xylem of stem

Soil surface

Root hairs

Epidermis

Cortex

Pericycle

Endodermis

Conducting xylem of root

sult, water exits the cells, thereby decreasing the cell's turgor and causing the stoma to close

e. This mechanism of potassium movement partially explains stomatal opening and closing; because the stomata of some plants open before daylight, another mechanism (yet to be discovered) must also be involved

III. Water Movement

A. General information

1. Water moves through a plant from root to stem and leaves through the xylem tissue in a process called *xylem transport* (see *Transpiration Pathway*)
2. Water (and dissolved minerals) from the soil that enter the xylem are filtered as they pass through the root cortex
3. The mechanism by which water moves through a plant involves root pressure and transpiration, but it also relies upon the unique physical characteristics of water

B. Cohesion-adhesion mechanism

1. Water molecules possess attractive forces that allow them to stick to one another (*cohesion*) or to various environmental surfaces (*adhesion*); these forces result from the arrangement of the hydrogen and oxygen atoms that make up the water molecule
2. Plants use solar energy and the cohesive properties of water to move water throughout the plant
   a. As sunlight heats the leaf surface, water evaporates from the leaf
   b. Water that evaporates from cell surfaces in the leaf interior is replaced by water from within the cell
   c. As the cell loses water, water diffuses from adjacent cells to replenish it
   d. When the movement of water between adjacent cells reaches the xylem cells within the leaf vein, it exerts a pulling force (or tension) on the water in the xylem
   e. Because of the cohesiveness of water molecules, this upward-pulling force is transmitted along the column of water within the stem xylem into the roots
   f. As water moves upward, the roots absorb more water from the surrounding soil
3. Xylem transport is largely solar powered, because most of the energy needed to move water from root to shoot results from the evaporation (transpiration) of water from the leaf and stem surfaces

IV. Translocation

A. General information

1. *Translocation* is the movement of food, minerals, and water within a plant
2. Materials that are translocated include sugar, amino acids, proteins, calcium, sulfur, phosphate, hormones, and ions
3. Translocation usually occurs within the phloem

B. Phloem function

1. Materials in phloem move from a source to a sink
   a. A *source* is an area in which translocated material is synthesized; for example, leaves undergoing photosynthesis are a source of carbohydrate (sugar)
   b. A *sink* is an area in which translocated material is either consumed or stored; roots, fruits, and flowers are examples of sinks
2. Sources and sinks change over time, depending on the status of the plant
   a. Active growth, fruit production, and storage of materials for winter affects sources and sinks
   b. In trees such as the sugar maple, sugars are translocated during the growing season from leaves (sources) to roots (sinks) for winter storage; in the early spring, when the trees produce new leaves, sugar is translocated from roots to shoots to supply energy for rapid growth
3. Materials are moved by active transport into the phloem in a source and out of the phloem in a sink
   a. Once the material to be translocated is deposited in phloem cells, water diffuses passively into these cells and builds up turgor pressure
   b. The turgor pressure of one phloem cell forces materials into adjacent cells
   c. The rate of movement within phloem is approximately 50 to 100 cm/hour

C. Phloem tissue

1. The transport tissue of phloem consists of sieve tube cells and companion cells
2. *Sieve tube cells* are long, narrow cells that serve as conduits for the translocated materials
   a. Their cell walls have characteristic perforations (resembling a sieve)
   b. These cells are specifically involved in transport
3. *Companion cells* are found adjacent to sieve tube cells
   a. They appear to play an active role in "loading" sieve tube cells with the material to be translocated from sources
   b. Companion cells in leaves (an area where loading takes place) are larger and more prevalent than those in stems, which are strictly transport areas

Study Activities

1. Diagram the flow of water through the two transport pathways of the root.
2. Name the four physical factors and two biologic factors that affect transpiration.
3. Discuss the role of the guard cells in opening and closing the stoma.
4. Describe the cohesion-adhesion mechanism of transpiration.
5. Explain the function of translocation and describe how it occurs.

8

Taxonomy and Classification

Objectives

After studying this chapter, the reader should be able to:
• Describe taxonomy in terms of its purpose and scheme.
• Explain the difference between phenetic and phylogenetic classification.
• Define binomial nomenclature and explain its importance.
• Name and describe the five kingdoms.

I. Taxonomy and Classification

A. General information

1. *Taxonomy* is the branch of biology involved in naming and classifying the various organisms
2. It is based on the differences between prokaryotic and eukaryotic cells
   a. *Prokaryotic cells* have no distinct nucleus or nuclear membrane and no organelles (internal structures with surrounding membranes); representative organisms are bacteria and cyanobacteria (blue-green algae)
   b. *Eukaryotic cells* have a distinct nucleus and nuclear membrane and numerous organelles; representative organisms are all those other than bacteria and cyanobacteria
3. Classification is based on similarities in structure, function, chemistry, behavior, habitat, reproduction, and development
4. Taxonomists may not always agree on the classification of specific organisms

B. Linnaean classification

1. The practice of classification consists of several steps: recognizing and describing related groups of organisms, fitting a newly recognized group into a formal hierarchy of classification, and assigning names to new organisms
2. Carolus Linnaeus, an 18th-century Swedish naturalist, devised the Linnaean classification scheme, which uses binomial nomenclature
   a. In *binomial nomenclature*, a naming scheme still used by modern biologists, each organism is assigned a unique two-part name
   b. The organism is identified by its genus and species; for example, the red maple is named *Acer rubrum* (Acer is the genus and rubrum is the species)
   c. Binomial nomenclature assigns latinized names to each organism because Latin is a "dead" language that is unlikely to change over time; Latin is also

the root of many modern languages, so latinized names transcend many
language barriers
   d. The scientific (genus and species) name is important because it can be recognized by all scientists; a common name may vary from region to region
   or apply to more than one species
   e. Names of organisms are established by a rigid set of rules and are adopted
   by international commissions of taxonomists

C. Hierarchical classification
   1. Organisms are classified in a hierarchical fashion, from kingdoms to species
   2. A *kingdom* is the broadest taxonomic group
      a. Organisms in this group do not share many traits
      b. The kingdom is subdivided into phyla (used in zoology) or divisions (used in
      botany)
      c. Phyla and divisions are subdivided into classes, orders, families, genus, and
      species
   3. A **species** is the smallest taxonomic group
      a. Organisms in this group are very similar
      b. The primary factor for determining species identity is reproductive isolation
      from other taxonomic groups; that is, members of the same biologic spe-
      cies can interbreed and produce fertile offspring
      c. Currently 1.4 million species have been recognized, named, and described
      d. Occasionally classification groups that are more narrow or specific than spe-
      cies, such as races, varieties, or subspecies, may be used

D. Phenetic versus phylogenic classification
   1. Organisms are classified based on physical structure and patterns of descent
   2. Phenetic classification is the oldest scheme and is based on similarity of appear-
      ance, structure, or function; no ancestral relationships are assumed so taxo-
      nomic groups of mixed ancestry are possible
   3. Phylogenic classification schemes reflect patterns of descent and taxonomic
      groups are defined by common ancestry

II. Classification of Kingdoms

A. General information
   1. Five kingdoms of organisms are currently recognized: Monera, Protista, Fungi,
      Plantae, and Animalia
   2. This classification scheme was proposed by Robert H. Whittaker, an American bi-
      ologist, in 1959

B. Kingdom Monera
   1. This kingdom comprises prokaryotic, single-cell (occasionally colonial or filamen-
      tous) organisms
   2. Bacteria and cyanobacteria (blue-green algae) are members of this kingdom
      a. Bacteria are autotrophic (chemosynthetic and photosynthetic) and heterotro-
      phic (saprobic and parasitic)
      b. Cyanobacteria are photosynthetic and can fix nitrogen

C. Kingdom Protista

1. This kingdom encompasses a diverse group of eukaryotic, unicellular, colonial, or filamentous organisms that may or may not have cell walls
2. Protists are generally small and inconspicuous; they may be photosynthetic, heterotrophic, or both
3. Examples include algae (plantlike protists, excluding cyanobacteria) and protozoa (animal-like protists), such as amoeba, diatoms, seaweed, slime molds, and ciliates

D. Kingdom Fungi

1. The kingdom Fungi consists of eukaryotic, filamentous organisms with **chitin** in their cell wall
2. These heterotrophic (saprobic and parasitic) organisms absorb nutrients from their surroundings
3. Mushrooms, mildew, yeast, and molds are fungi

E. Kingdom Plantae

1. Multicellular eukaryotic, autotrophic, and photosynthetic organisms make up the kingdom Plantae
2. All plants contain chlorophyll a and b, and most live on land
3. The major subdivisions of the plant kingdom are Bryophyta and Tracheophyta
   a. *Bryophytes* are nonvascular, seedless plants; examples are mosses, liverworts, and their relatives
   b. *Tracheophytes* have vascular tissue for transporting water and nutrients; examples include seedless plants and seed plants
      (1) Seedless vascular plants, which reproduce by spores, include ferns and their relatives
      (2) Seed plants include conifers and evergreens (gymnosperms) and flowering plants (angiosperms)

F. Kingdom Animalia

1. This kingdom consists of eukaryotic, heterotrophic, multicellular organisms
2. The major phyla of the animal kingdom are Porifera (sponges); Cnidaria (jellyfishes, corals, and sea anemones); Platyhelminthes (flatworms, flukes, and tapeworms); Nemertina or Aschelminthes (roundworms); Annelida (segmented worms, leeches, and earthworms); Arthropoda (animals with jointed appendages and segmented bodies, such as insects, crayfish, spiders, centipedes, horseshoe crabs, and barnacles); Echinodermata (starfish, sea urchins, and sand dollars); and Chordata (animals with a spinal cord, such as fish, amphibians, reptiles, birds, and mammals)

Study Activities

1. Explain the binomial system of classification.
2. Name the seven major taxonomic groups.
3. Describe the organisms in each of the five kingdoms and give examples of each.

9

Viruses and Kingdoms Monera and Protista

Objectives

After studying this chapter, the reader should be able to:
- Describe the features of viruses.
- Identify the major characteristics of the kingdom Monera.
- Discuss the ecologic importance of bacteria.
- Identify the major taxonomic groups of the kingdom Protista.
- Explain the life cycles of representative protists.
- Name the economic products derived from algae.
- Explain the cause and consequences of red tides.

I. Viruses

A. General information
    1. Viruses are nonliving entities that consist of strands of nucleic acid surrounded by an exterior protein coat
    2. Because viruses lack ribosomes and the enzymes necessary for protein synthesis, they can survive and reproduce only in living host cells
    3. Viruses infect virtually all forms of life, but they are generally host-specific, infecting only one type of host cell

B. Infection
    1. Viruses cause a number of diseases in animals and plants
    2. Viral diseases in plants are usually transmitted by insects (such as aphids, leafhoppers, and scale insects) or other invertebrates
    3. Common symptoms of viral infection in plants are necrosis (patches of dead tissue), spotting of leaves or stems, abnormal color patterns, and tumor formation
    4. Most plant viruses contain ribonucleic acid (RNA), rather than deoxyribonucleic acid (DNA), as the genetic material
    5. The proteins and nucleic acids of viruses change rapidly, continually producing new strains, which can make identification and treatment difficult

II. Kingdom Monera

A. General information
    1. The kingdom Monera comprises bacteria and cyanobacteria (blue-green algae)

2. Although both viruses and bacteria can cause infections, these organisms are not closely related

3. All organisms in the kingdom Monera are prokaryotic cells, the oldest and structurally the simplest forms of life

   a. The prokaryotic, or bacterial, chromosome consists of a single circular structure

   b. The DNA in bacterial chromosomes is simpler and has fewer associated proteins than in eukaryotic chromosomes

   c. The bacterial chromosome is located in a specific area of the cytoplasm known as the nucleoid region, but it is not separated from the rest of the cell

   d. Many bacteria also have *plasmids*, small rings of DNA that carry accessory genes and often function and replicate independently of the larger bacterial chromosome

B. Characteristics

1. Monera can be characterized by their size and shape, method of reproduction, presence of cell walls, response to *Gram stain*, and ability to form capsules

2. Prokaryotic cells lack the extensive internal membranes and structures characteristic of eukaryotic cells

3. Almost all members of this kingdom are tiny, single-cell organisms with no true cellular specialization

   a. Some cyanobacteria consist of a filament with two or three specialized cell types

   b. Bacteria are classified by shape: cocci (spheres), bacilli (rods) and spirochetes (spirals)

4. As prokaryotic organisms, Monera do not use mitosis or meiosis as a means of cell division; reproduction is primarily asexual by means of fission

5. All Monera have cell walls, which maintain the shape of the cell and provide physical protection

   a. Instead of the cellulose that makes up plant cell walls, bacterial cell walls contain large molecules called peptidoglycans

   b. *Peptidoglycans* are large polymers of modified sugars cross-linked with peptides (amino acid residues)

   c. Bacteria differ in the amount of peptidoglycan found in their cell walls

   d. Many antibiotics, such as penicillins, inhibit the formation of the peptide cross-links with peptidoglycan, thus preventing the bacteria from constructing a functional cell wall

   e. Antibiotics can therefore halt the growth of many bacterial species without adversely affecting eukaryotic cells

6. Bacteria can be identified by Gram stain, which is named for the Danish physician Hans Christian Gram, who developed the technique in the 1800s

   a. In the Gram stain test, bacteria are stained with violet dye and iodine, rinsed in alcohol, and stained again with red dye

      (1) The violet dye stains the peptidoglycan layer of the cell wall

      (2) The red dye stains lipopolysaccharides (molecules containing lipids and sugars)

   b. Gram-positive bacteria have cell walls that contain large amounts of peptidoglycan, which retains the violet dye

   c. Gram-negative bacteria have cell walls that contain smaller amounts of pepti-
      doglycan; when the violet dye is washed from the cell wall by the alcohol,
      the cell walls take up the red dye
   d. Gram-positive bacteria are susceptible to antibiotics; gram-negative bacteria
      are often more resistant to antibiotics because of their outer layers of
      lipopolysaccharides
7. Many prokaryotes secrete sticky substances that form a capsule on the outside
   of the cell wall
   a. Capsules provide the cell with additional protection
   b. They also enable the cell to adhere to a substrate or other adjacent cells

C. Importance
1. Bacteria have both helpful and harmful functions
2. They are *decomposers*, organisms that survive on dead organisms
   a. In this role, bacteria are natural recyclers within ecosystems
   b. They release nutrients from decaying organisms to be used again by living
      organisms
3. Bacteria are used in industrial processes of fermentation to produce alcohol and
   other organic products
4. They are the predominate organisms used in the genetic engineering of pharmaceutical and industrial products
5. Bacteria can also cause many human diseases, such as cholera, leprosy, and
   tetanus, and are responsible for general infections of wounds and body organs
6. Bacteria are also important plant pathogens
   a. Almost all plants can be affected by diseases caused by plant pathogenic
      bacteria
   b. Approximately 12% of crops worldwide are lost to bacterial disease
   c. The majority of plant pathogenic bacteria are bacilli that reside in host plants
      as parasites
   d. These bacteria cause various symptoms that commonly include wilting, soft
      rot, and blight
      (1) Wilting results when bacterial cells invade the xylem vessel cells, inter-
          fering with the movement of water and inorganic nutrients in the plant
      (2) Soft rot most commonly occurs on fleshy storage organs of vegetables,
          such as potatoes and onions, or fleshy fruits, such as tomatoes and
          eggplants
      (3) Blight is characterized by the rapid development of dead, discolored
          areas on stems, leaves, and flowers

D. Physiology
1. Bacteria use diverse metabolic processes for obtaining energy and producing
   substances necessary for survival
2. *Saprobes* obtain nutrients from dead organic matter
3. Photosynthetic bacteria obtain energy through photosynthesis
4. Chemoautotrophs derive energy from the oxidation of inorganic molecules, such
   as hydrogen sulfide, ammonia, or ferrous ions
5. Nitrogen-fixing bacteria can convert atmospheric nitrogen gas (N_2) to nitrogenous
   forms (ammonia) that can directly or indirectly be used by plants or other organisms

E. Adaptive behavior
1. To survive during periods of environmental stress, many species of bacteria form endospores
2. **Endospores** are structures consisting of dormant cells surrounded by a tough outer capsule
3. Endospores are resistant to desiccation (drying) and heat, which allows survival under adverse conditions

F. Reproduction
1. Bacteria reproduce through binary fission, a process in which the cell wall grows inward, forming two new separate cells or two cells that remain joined to form chains
2. The genetic material of bacteria is a circular loop of DNA located within the cytoplasm
   a. During bacterial reproduction, the single circular loop replicates to form two loops
   b. One identical DNA loop enters each of the newly formed bacterial cells

G. Taxonomic groups
1. Members of the kingdom Monera are divided into two taxonomic groups: Archaebacteria and Eubacteria
2. *Archaebacteria* are the earliest prokaryotes
   a. They are found in nearly all habitats, including the most extreme—swamps, marshes, intestinal tracts of animals, salt lakes, hot springs, and sulfur springs
   b. Archaebacteria differ from Eubacteria in several ways
      (1) Their cell walls lack peptidoglycan, a chief component of eubacteria cell walls
      (2) Their plasma or cell membrane contains lipids, unlike those of any other organisms
      (3) Their ribosomal protein and RNA polymerase are similar to those found in eukaryotic cells but very different from those in Eubacteria
3. *Eubacteria*, or true bacteria, are the largest group of bacteria
   a. These prokaryotes display great metabolic diversity
   b. They (along with fungi) are responsible for decomposing and recycling of all types of organic matter
   c. Some species are parasitic (use living organisms for food), and many cause human disease
   d. Other species, such as those in the genera *Rhizobium, Azotobacter,* and *Clostridium,* can fix nitrogen
   e. The nitrogen-fixing capabilities of *Rhizobium* allow this bacteria to develop a symbiotic (mutually beneficial) relationship with a plant
      (1) *Rhizobium* typically infects the roots of plants in the legume family (which includes peas, beans, soybeans, peanuts, alfalfa, and clover)
      (2) The infection produces root nodules (swellings) that are composed of plant cells containing the bacteria
      (3) The bacteria within the nodules, called bacteroids, supply the plant with nitrogen; the plant supplies the bacteria with carbohydrates and other organic compounds

   (4) In the agricultural practice known as crop rotation, legumes are often planted and plowed under at the end of the growing season to increase soil nitrogen

 4. Cyanobacteria are a subgroup of Eubacteria (see *Representative Cyanobacteria* for an illustration)

  a. Cyanobacteria (blue-green algae) can obtain energy through photosynthetic processes similar to those of green plants; cyanobacteria contain chlorophyll a and produce oxygen as a byproduct

   (1) Cyanobacteria may possess carotenoids (red and yellow pigments) or phycobilin (blue pigments) in addition to chlorophyll a

   (2) Combinations of these pigments give some cyanobacteria a dark (brown or black) appearance

  b. Some genera (*Anabaena* and *Nostoc*) are capable of nitrogen fixation

  c. Cyanobacteria are commonly found in water, soil, and other moist surfaces

   (1) In polluted waters, these organisms may produce algae blooms (rapid population explosions)

   (2) Algae blooms destroy the aesthetic appearance of the water and may rob the water of oxygen as the cyanobacteria die and decay, thereby suffocating aquatic organisms

III. Kingdom Protista

A. General information

 1. The kingdom Protista includes both animal-like and plantlike organisms

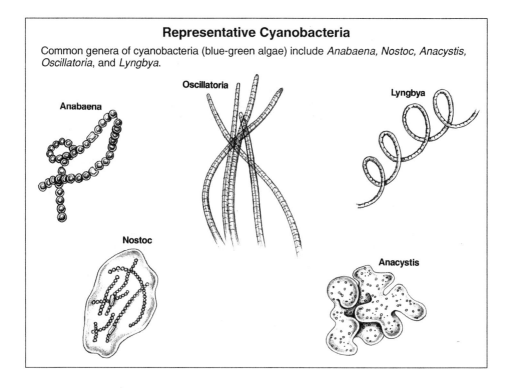

Representative Cyanobacteria

Common genera of cyanobacteria (blue-green algae) include *Anabaena, Nostoc, Anacystis, Oscillatoria,* and *Lyngbya.*

Oscillatoria

Lyngbya

Anabaena

Nostoc

Anacystis

2. Historically, scientists had classified these organisms as members of the Plant or Animal kingdom
3. Protists can be single-celled, multicellular, or colonial; all are eukaryotes, although most have little cellular specialization
4. These organisms serve as the basis of many freshwater and marine food chains; as photosynthetic organisms, they are the route by which solar energy enters aquatic ecosystems
5. Animal-like protists are called protozoa, and plantlike protists are called algae
   a. Protozoa are classified according to nutrition, mode of locomotion, and cell wall characteristics
   b. Algae are classified according to pigmentation, type of storage reserve, and cell wall characteristics
      (1) Green algae, red algae, and dinoflagellates store energy (food) reserves as starch
      (2) Brown algae store energy reserves as *laminarin*, which is a glucose-containing polysaccharide that resembles starch, but has different linkages between the glucose molecules
      (3) Diatoms store energy as *chrysolaminarin*, which is a more highly polymerized form of laminarin

B. Life cycles

1. Protists reproduce through sexual reproduction
2. Like other living organisms, protists exhibit one of three sexual life cycles: haplontic, alternation of generations, and diplontic (see *Generalized Life Cycles*, page 74)
3. In the *haplontic life cycle*, adults are haploid organisms
   a. Mitosis produces haploid gametes that fuse to produce a diploid zygote
   b. The zygote undergoes meiosis to form haploid cells, which subsequently divide by mitosis
   c. The haploid cells continue to increase in number by mitosis and may remain as single cells or form multicellular organisms
   d. The haplontic life cycle is common in algae and fungi and is exemplified by the algae *Chlamydomonas*
4. In the *alternation of generations*, a haploid generation alternates with a diploid generation
   a. Diploid organisms (called sporophytes) produces haploid spores as a result of meiosis
   b. The spores do not function as gametes, but undergo mitosis to give rise to multicellular haploid organisms (called gametophytes)
   c. The gametophytes eventually produce gametes that fuse to form diploid zygotes, which, in turn, differentiate into diploid organisms
   d. Alternation of generations is characteristic of plants and many algae, particularly brown algae (except *Fucus*)
5. In the *diplontic life cycle*, adults are multicellular diploid organisms
   a. Meiosis produces haploid gametes in the diploid adult
   b. The gametes fuse to form a diploid zygote, which divides to produce another multicellular diploid organism
   c. The diplontic cycle is characteristic of animals and some protists, such as the brown algae *Fucus*

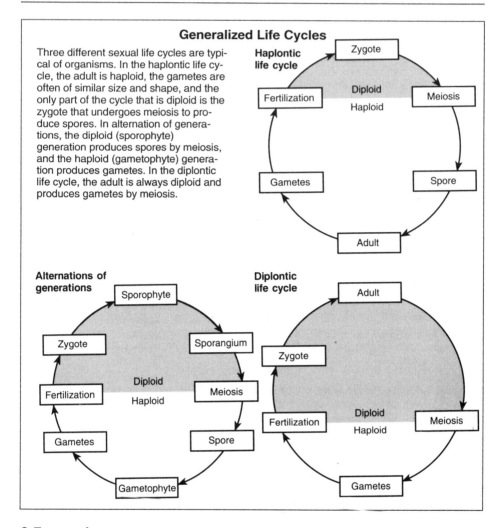

Generalized Life Cycles

Three different sexual life cycles are typical of organisms. In the haplontic life cycle, the adult is haploid, the gametes are often of similar size and shape, and the only part of the cycle that is diploid is the zygote that undergoes meiosis to produce spores. In alternation of generations, the diploid (sporophyte) generation produces spores by meiosis, and the haploid (gametophyte) generation produces gametes. In the diplontic life cycle, the adult is always diploid and produces gametes by meiosis.

Haplontic life cycle

Alternations of generations

Diplontic life cycle

C. Taxonomic groups

1. The plantlike, or algal, protists comprise the phyla Chrysophyta, Chlorophyta, Phaeophyta, Rhodophyta, Pyrrhophyta, and Euglenophyta (see *Representative Algal Protists* for an illustration); animal-like protists include Rhizopoda, Sporozoa, Ciliophora, and Zoomastigina

2. *Chrysophyta* (golden brown algae and diatoms) are photosynthetic, unicellular organisms that live in fresh and salt water and moist soils
    a. They possess chlorophylls a, b, and c and the pigments carotenoid and fucoxanthin
    b. Many varieties have an outer double shell of silica, which resembles microscopic glass bottles; this outer shell has grooves and pores to allow the exchange of gases, nutrients, and wastes between the cell and its environment
        (1) Diatomaceous earth is formed from the external glassy shells of previous generations of diatoms that have accumulated in ancient oceans

Representative Algal Protists

The kingdom Protista contains a variety of organisms, including the common algal protists depicted here.

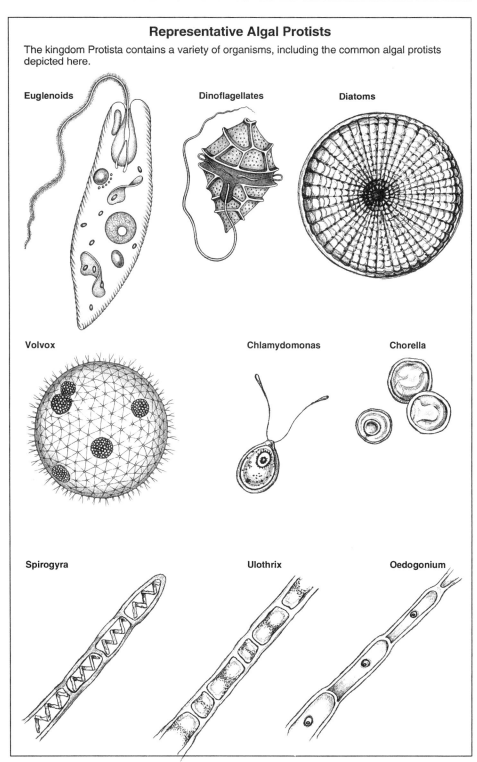

Euglenoids

Dinoflagellates

Diatoms

Volvox

Chlamydomonas

Chorella

Spirogyra

Ulothrix

Oedogonium

Illustrations of Euglenoids, Dinoflagellates, and Volvox from Zwolski, Kenneth. Applied Science Review: *Biology*, Springhouse, Pa.: Springhouse Corp.,1993.

   (2) These deposits are mined as a powder that is used in filtration systems (swimming pool filters and municipal water treatment plants), as a light abrasive (toothpaste and polishes), and in the light-reflecting paints used on roadways and signs
  c. Others varieties lack cell walls or have walls composed of cellulose impregnated with minerals
  d. Most golden brown algae and diatoms store carbohydrate in the form of chrysolaminarin and lipids rather than starch
  e. Diatoms generally reproduce asexually by mitosis, but there is some evidence of sexual reproduction
   (1) Diatom cells are usually *diploid* (having two complete sets of chromosomes in the nucleus)
   (2) When the diploid cells experience favorable conditions or reach a critical size, they undergo mitosis to form two daughter cells that separate and become individual organisms
   (3) Each daughter cell produces a new cell wall that fits inside the old wall; as a result, the cells become progressively smaller
   (4) These smaller cells ultimately undergo meiosis to form gametes (reproductive cells), which are released into the environment
   (5) The gametes unite to form a zygote (called an auxospore) that dramatically increases in size; when mature, the auxospore forms a new external rigid wall of silica, thereby restoring the organism to its original size
3. Chlorophyta (green algae), which are thought to be the direct ancestors of plants, share a number of characteristics with modern plants
  a. They possess chlorophylls a and b and accessory pigments, use starch as a storage carbohydrate, and have a cell wall composed of cellulose
  b. Like Chrysophyta, Chlorophyta are found in fresh and salt water and moist soils
  c. These organism may be unicellular, filamentous, or multicellular
  d. Common green algae include *Chlorella, Chlamydomonas, Oedogonium, Ulothrix, Volvox, Spirogyra,* and *Ulva*
  e. The sexual life cycle of multicellular, filamentous green algae is typified by *Ulothrix*
   (1) **Asexual reproduction** begins with the release of a **zoospore** (a motile reproductive cell that can develop directly into a new organism and does not unite with other cells) from the parent cells of a mature filament
   (2) The zoospore develops a new filament by repeated cell division (mitosis) and forms a holdfast, which attaches the filament to a substrate
   (3) The cytoplasm of any cell except the holdfast can clump and condense inside the cell wall, divide by mitosis, and become zoospores
   (4) The zoospores escape from the parent cell through a pore in the cell wall, which completes the asexual cycle
   (5) Sexual reproduction begins when the cytoplasm of a mature filament condenses and subsequently divides by mitosis to produce gametes that are released from the parent cell
   (6) The gametes possess two flagella each and are identical in size and appearance (sexual reproduction involving gametes of equal size and appearance is called **isogamy**)

(7) The gametes fuse to form a zygote, which eventually settles onto the substrate, undergoes a period of dormancy, and eventually produces zoospores through meiosis

(8) The zoospores escape from the thick walls of the zygote, settle on a substrate, and develop a new filament through mitosis

(9) The zygote is the only portion of the life cycle that is diploid, the rest of the life cycle is haploid

f. The sexual life cycle of unicellular green algae is typified by *Chlamydomonas*

(1) Asexual reproduction begins with the loss of the parent cell's two flagella and continues with the division of the nucleus by mitosis, resulting in two cells within the cell walls of the parent cell

(2) Each daughter cell develops flagella and escapes as the parent cell wall breaks down

(3) In some cases, multiple mitotic divisions occur within the same parent cell, producing a cluster of daughter cells that may remain together as a colony until growth conditions change

(4) All cells in the asexual cycle of *Chlamydomonas* are haploid

(5) Under certain environmental conditions (often those involving environmental stress, such as starvation), *Chlamydomonas* undergo sexual reproduction

(6) Sexual reproduction begins when the parent cell loses its flagella and continues with the production of numerous gametes by repeated mitotic divisions

(7) The gametes released from the parent cell fuse with gametes from different *Chlamydomonas* strains to form a diploid zygote

(8) After a period of dormancy, the diploid zygote undergoes meiosis to produce four haploid cells known as zoospores

(9) When the zygote wall breaks down, the zoospores are released, swim away, and mature into full-sized *Chlamydomonas* cells

4. Phaeophyta (brown algae) are multicellular organisms that primarily inhabit sea water

a. Brown algae are named for the coloration given them by the brown pigment fucoxanthin, which is abundant in their cells

b. These organisms also contain chlorophylls a and c in their chloroplasts

c. They store food in the form of laminarin (a polymer of glucose) and mannitol (an alcohol)

d. Phaeophyta include many seaweeds and kelps (genera *Sargassum, Fucus, Nereocystis*) that dominate the rocky shores along coastlines of cooler regions of the world, but also are found in extensive offshore beds

e. They are economically important sources of salt, iodine, and algin—a gelatinous polysaccharide found in the cell wall, which is extracted from kelp and used to stabilize products and enhance their creamy texture (for example, algin is added to ice cream, salad dressing, jelly beans, toothpaste, pharmaceuticals, milkshakes, latex paints, paper coatings, and polishes)

f. Brown algae range in size from microscopic to the largest seaweeds, which are differentiated into the holdfast, stipe, and blade

(1) The *holdfast* is a tough, strong structure that superficially resembles a mass of roots and anchors the seaweed to the substrate

(2) The *stipe* is a hollow stalk structure that supports the upper portions of the seaweed and contains growth centers that allow the seaweed to increase in height or length

(3) The *blade* makes up the rest of the body of the seaweed and is responsible for most of the organism's photosynthesis

g. The sexual life cycle of brown algae is typified by *Fucus*, or rockweed

(1) Rockweed grows as separate male and female "plants" called thalli (the term *thallus* is used to describe a flattened, multicellular plant body not organized into leaves, stems, and roots)

(2) At the tips of the thalli are conceptacles (swellings) that contain **antheridia** (male reproductive structures) or **oogonia** (female reproductive structures)

(3) Oogonia produce eggs; antheridia produce sperm

(4) Eggs and sperm released from the oogonia and antheridia unite to form diploid zygotes, which grow to form new thalli

(5) The newly formed thalli are diploid and grow new conceptacles to complete the cycle

5. Rhodophyta (red algae) are multicellular organisms most often found in warm ocean waters, although many inhabit cooler waters; they are the coastal seaweeds of intertidal areas

a. Red algae are named for the coloration given them by the accessory pigments phycobilins

(1) Phycobilins mask the presence of chlorophyll a

(2) These accessory pigments enable red algae to absorb the green, violet, and blue light that penetrates deep water

b. Red algae store food in the form of starch

c. These organisms generally exist as filamentous forms, although the filaments may be so tightly coiled that they appear as flattened blades

d. Red algae usually grow attached to rocks or other algae, but some are free-floating and a few are colonial or unicellular

e. They are economically important sources of agar and carrageenan

(1) Agar and carrageenan are sulfur-containing polysaccharides that are components of the algal cell wall

(2) Agar is a gelatin used in laboratories and medical testing facilities as a culture medium for growing bacteria; it is also used in the manufacture of capsules for drugs and is added to baked goods to retain moistness

(3) Carrageenan is an extracted from a red algae called Irish moss and is used as a thickener and stabilizer in paints and dairy products

6. Pyrrhophyta (dinoflagellates) are primarily unicellular marine organisms, although some freshwater forms have been identified

a. Dinoflagellates contain chlorophylls a and c and carotenoids

b. The majority of dinoflagellates are photosynthetic, but approximately 45% are nonphotosynthetic heterotrophs

c. These organisms store energy in the form of starch

d. They have two flagella (whiplike structures protruding from individual cells), which are used for locomotion; the external surface of the cells is composed of several cellulose plates

e. Some forms, known as zooxanthellae, develop symbiotic relationships with such animals as sea anemones, mollusks, and corals

f. Some Pyrrhophyta species experience periodic population explosions, or "blooms," which cause red tides
   (1) The blooms produce toxins that are harmful to many organisms and often accumulate in shellfish at levels poisonous to humans
   (2) When the algae bloom dies, it often depletes the water's oxygen
7. Euglenophyta (euglenoids) are protists that possess characteristics of both plant and animal cells
   a. They are similar to animals because they lack a cell wall, are capable of movement, and can obtain food by heterotrophic means
   b. They are similar to plants because their cells contain chloroplasts for photosynthesis
   c. Little is known about their reproductive cycle, although most reproduction is thought to occur asexually by cell division
   d. Euglenoids store energy reserves as paramylum, a unique storage carbohydrate
   e. The single-cell, flagellated protist *Euglena* is a member of this group
8. Rhizopoda (amoebas) are heterotrophic, animal-like protists found in fresh water, salt water, and moist soils; they lack cell walls and move via cytoplasmic projections called pseudopodia
9. Sporozoa are nonmotile, spore-forming, animal parasites responsible for some human diseases; *Plasmodium*, for example, causes malaria
10. Ciliophora (ciliates) are unicellular, heterotrophic protists; the paramecium belongs to this group
11. Zoomastigina are unicellular, heterotrophic organisms, some of which are associated with human diseases; *Trypanosoma,* for example, causes sleeping sickness

Study Activities

1. Name five common symptoms caused by plant viruses.
2. List the major characteristics of the kingdom Monera.
3. Explain why bacteria are ecologically important.
4. Describe the symbiotic relationship between *Rhizobium* and legume roots.
5. List the major features of the phyla Chrysophyta, Chlorophyta, Phaeophyta, Rhodophyta, Pyrrhophyta, and Euglenophyta.
6. Name three economic products obtained from algae.

10

Kingdom Fungi

Objectives

After studying this chapter, the reader should be able to:
• Describe the major features of the kingdom Fungi and its three subkingdoms.
• List distinguishing characteristics for each major group of fungi.
• Trace the life cycles of representative fungi.
• Describe the ecologic and economic importance of fungi.
• Discuss the mutualistic relationships formed by lichens and mycorrhizae.

I. Fungal Structure, Nutrition, and Ecology

A. General information

1. Fungi historically have been classified with plants, but major differences between these organisms necessitated the formation of a separate kingdom
2. Fungi are eukaryotic, usually multicellular, multinucleate organisms
3. They obtain nutrients by secreting enzymes into their substrate and absorbing the digested materials
4. Most fungal cell walls contain the polysaccharide chitin
5. The fungal structure is characterized by long slender filaments called **hyphae** (see *Types of Fungal Hyphae* for an illustration)
   a. Coenocytic hyphae consist of long cellular strands with many nuclei, which are not contained within individual cells
   b. Hyphae with complete septa (barriers) are separated into individual cells; septate hyphae may contain a single nucleus in each cell or two nuclei in each cell (called dikaryotic hyphae)
   c. Hyphae with incomplete septa allow cytoplasm to flow freely from cell to cell
   d. Some parasitic fungi have **haustoria**, which are specialized hyphae that extend into individual host cells
6. The **mycelium** is the mass of hyphae that constitutes the body of a fungus; it is the most recognizable structure of a mushroom (toadstool)

B. Modes of nutrition and metabolism

1. Fungi (along with bacteria) are the principal decomposers in every ecosystem
2. They can break down lignin, a major component of wood
3. Some fungi also attack living organic matter, causing agricultural damage and destroying food stores; food contaminated by fungi may be unpalatable or poisonous to humans

Types of Fungal Hyphae

This illustration depicts various types of fungal hyphae.

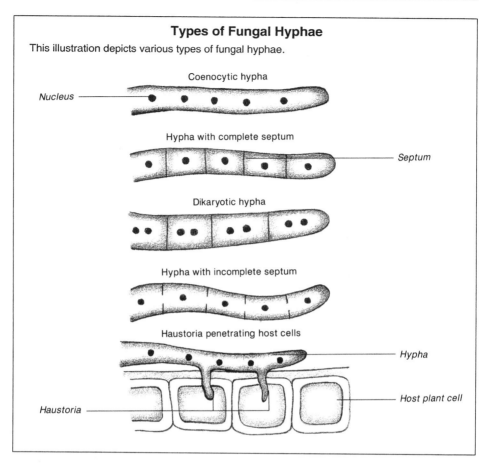

4. Warm, moist, dark conditions are most favorable for fungal growth; however, fungi can grow in various other habitats
5. If conditions become too stressful, fungi survive by producing spores, which are resistant to temperature and moisture extremes
6. Parasitic fungi use haustoria to penetrate individual host cells and absorb nutrients directly from the cytoplasm

C. Ecology

1. Fungi typically form **symbiotic** relationships with plants or algae
2. Two types of fungal symbionts are lichens and mycorrhizae
3. **Lichens** are a combination of fungi (Ascomycetes or Basidiomycetes) and green algae (phylum Chlorophyta) or cyanobacteria (kingdom Monera)
   a. The algae portion of the lichen provides food through photosynthesis; the fungal portion acts as a living sponge to improve water retention
      (1) Some evidence suggests that the fungal partner parasitizes the algae in a controlled fashion and sometimes may even destroy algal cells
      (2) The fungal component of lichens rarely grows independently, although the algal component may do so

b. Lichens inhabit cold, dry, and generally harsh environments, where they help break rock surfaces and prepare the habitat for other organisms

c. They survive harsh or adverse conditions by becoming dehydrated, which subsequently slows their metabolism

(1) When the lichen's water content drops dramatically, the upper portion of the thallus becomes opaque enough to exclude light from the photosynthetic algae

(2) Dehydrated lichen are unaffected by most environmental extremes because they are temporarily dormant and do not engage in photosynthesis

d. Lichens absorb nutrients from rain and air; for this reason, they are very sensitive to air quality and are among the first organisms to perish in a polluted environment

e. Lichens are grouped into three major growth forms

(1) *Crustose lichens* attach to or embed in their substrate and often form brightly colored, crusty patches on bare rocks and tree bark

(2) *Foliose lichens* have leaflike thalli, are weakly attached to their substrate, and have edges that are crinkly or divided into lobes

(3) *Fruticose lichens* resemble miniature upright shrubs or may hang from tree branches; their thalli are usually branched and cylindrical

4. **Mycorrhizae** are a combination of fungi and plant roots

a. Mycorrhizae enhance the absorption of essential nutrients by plant roots

(1) They also may provide protection against the effects of acidic soil and may make a plant more resistant to drought, cold, and harsh conditions

(2) They can prevent the accumulation of toxic metals in plants

(3) They can help plants to grow better in poor soils

(4) They can speed the seed germination of orchids

b. About 90% of all plants have a symbiotic relationship with mycorrhizae

c. Plants with mycorrhizae develop fewer root hairs than those without mycorrhizae; the mycorrhizae perform the same functions as root hairs, making them less necessary

d. Mycorrhizae are highly susceptible to acid rain, which may negatively affect the growth of some plants

e. The two forms of mycorrhizae are endomycorrhizae and ectomycorrhizae

(1) *Endomycorrhizae*, the more common variety, are characteristic of many crop species and develop when the fungal hyphae penetrate the outer root cells

(2) *Ectomycorrhizae*, the less common variety, are characteristic of shrubs and trees and develop when the hyphae surround, rather than penetrate, the root cells

II. Reproduction

A. General information

1. Fungi reproduce both asexually and sexually by means of spores

2. Fungi also can reproduce asexually by means of binary fission, budding, or fragmentation

B. Spore production
1. **Spores** are unicellular reproductive cells capable of developing into adult organisms without fusing with another cell
2. Spores enable fungi to colonize in new areas
3. Because they typically are resistant to temperature and moisture extremes, spores can survive harsh environmental conditions
4. Spores are produced by cell division during asexual or sexual reproduction
    a. Asexual spores are produced directly from the cells of the hyphae
    b. Sexual spores are produced in a saclike structure called the ascus
    c. Sporangiospores are produced in specialized hyphae called **sporangiophores** (spore-bearing hyphae), which contain specialized structures for spore formation (**sporangia**)
5. Some fungal species (Ascomycetes and Basidiomycetes) produce spores in a specialized reproductive structure called a fruiting body

C. Asexual reproduction
1. Fungi reproduce asexually by binary fission, budding, or fragmentation
2. In *binary fission*, individual fungal cells divide to form two identical daughter cells, which in turn grow into new individual organisms
3. In *budding*, which is usually restricted to single-cell fungi (such as yeast), a new organism is formed from a small, pinched-off portion of a mature cell
4. In **fragmentation**, strands of hyphae that are mechanically separated from the original mycelium grow and develop independently into new organisms

D. Sexual reproduction
1. Not all fungi reproduce sexually
2. Fungi do not have distinguishable male and female sexes; they have mating types, usually designated as "plus" and "minus" strains
3. Gametes are produced by specialized hyphae called **gametangia**
4. During fertilization, the two haploid gametes fuse to form a diploid zygote
5. After the zygote forms, the nucleus undergoes meiosis to produce new (now haploid) spores, which disperse, settle into favorable habitats, and grow to produce new organisms
6. The largest portion of the fungal life cycle is haploid; only the zygote is diploid

III. Taxonomic Classification

A. General information
1. Fungi are generally classified according to the characteristics of their spores and fruiting body
2. The kingdom Fungi is divided into three subkingdoms: Mastigomycotineae, Eumycotineae, and Myxomycotineae
3. Authorities do not agree on fungal classification; Myxomycotineae are sometimes considered part of the kingdom Protista

B. Subkingdom Mastigomycotineae (water molds)
1. Water molds share several features with brown algae, from which they are thought to have evolved
    a. Their cell walls are composed of cellulose

b. Their bodies vary from unicellular to highly branched coenocytic and filamentous forms

c. Their spores are flagellated and require free-standing water for swimming

2. Many water molds are aquatic, commonly growing on dead insects and plant debris in water

3. Mastigomycotineae reproduce sexually, which results in the formation of spores that are dispersed to form new mycelia

4. Examples include *Saprolegnia*, a parasite of fish that grows on cuts and bruises and is a common aquarium pest, and *Phytophthora infestans*, the species responsible for the potato blight that devastated Ireland in the mid-1800s

C. Subkingdom Eumycotineae (true fungi)

1. True fungi are filamentous organisms that have a cell wall containing chitin but do not possess motile cells

2. The four major phyla in this subkingdom are Zygomyocota, Ascomycota, Basidiomycota, and Deuteromycota

3. Zygomyocota (black bread molds) is a common fungi that grows on bread and other baked goods and is thus responsible for their spoilage

a. When spores land on a suitable substrate, they germinate and produce extensive coenocytic mycelia

b. Once the mycelia is established, asexual reproduction begins in the sporangiophores, which produce black spores

c. As the walls of the sporangia break down, the spores are released and carried away by air currents to germinate and establish the cycle again

d. Sexual reproduction occurs when the gametangia of two different mating strains fuse to form a zygospore

e. The zygospores may remain dormant for months, but eventually undergo meiosis and germination to produce spore-bearing hyphae

f. The spores produced in these hyphae are released to start the cycle again

4. Ascomycota (sac fungi) are the largest class of sexually reproducing fungi and include yeast, powdery mildews, molds, morels, and truffles

a. This class is named for the **ascus**, a reproductive structure in which spores are produced

b. The hyphae of sac fungi have incomplete septa dividing adjacent cells

c. Some ascomycetes cause plant diseases, including Dutch elm disease, chestnut blight, and ergot

d. Asexual reproduction is accomplished by means of spore production, budding, or binary fission

(1) Some ascomycetes produce long chains of asexual spores (**conidia**) in specialized hyphae called conidiophores

(2) Other ascomycetes (such as yeasts) reproduce asexually by binary fission and budding

e. Sexual reproduction occurs when the haploid hyphae of two different mating strains grow together

(1) The fusion of cells at the tip of mating haploid hyphae produces new, dikaryotic hyphae whose cells each contain two nuclei (one nucleus from each mating strain); the fusion of the cytoplasm of the hyphal cells is called **plasmogamy**

(2) When dikaryotic hyphae mature, the nuclei in some of the cells fuse to form an ascus; the fusion of the nuclei within the dikaryotic hyphae is called **karyogamy**

(3) Numerous asci form a layer of cells, or **ascocarp**, within a fruiting body

(4) Within individual asci, the diploid nuclei undergo meiosis and subsequent mitosis to form a total of eight haploid **ascospores** (spores developed from an ascus)

(5) The ascus ruptures to release the ascospores, which are dispersed to germinate and establish new hyphae

5. Basidiomycota (club fungi) include mushrooms, puffballs, bracket fungi, rusts, and smuts

   a. Asexual reproduction is not common among basidiomycetes; when it occurs, it is usually by means of spore production

   b. Sexual reproduction begins with the formation of spore-producing structures, called **basidia**, within the fruiting body

   (1) The spores (**basidiospores**) germinate to produce the primary mycelium, a haploid structure that contains mononucleate cells (those having only one nucleus)

   (2) Two compatible mating strains of primary mycelia grow together and fuse to form a secondary mycelium, which contains dikaryotic cells (those possessing two haploid nuclei—one from each mating strain)

   (3) The secondary mycelium grows to form a dense, solid mass called a button or young **basidiocarp**

   (4) The button expands to form the mature basidiocarp, which is the readily recognizable structure of a mushroom

   (5) The umbrella-like basidiocarp consists of thin gill-like structures that house the basidia

   (6) As each basidium matures, karyogamy (nuclear fusion) takes place and is immediately followed by meiosis, a process that produces the haploid basidiospores

   (7) The basidiospores are dispersed to germinate, form a primary mycelium, and complete the cycle

6. Deuteromycota (imperfect fungi) are so named because their life cycle is imperfect, that is, asexual

   a. As far as scientists know, imperfect fungi reproduce only asexually, usually by means of spore production (conidia)

   b. Deuteromycetes include *Penicillium notatum* (from which the antibiotic penicillin is extracted), *Penicillium roquefortii* (which gives flavor to Roquefort cheese), and *Aspergillus tamarrii* (which is used to produce soy sauce); other members of this phyla cause athlete's foot and ringworm

D. Subkingdom Myxomycotineae (slime molds)

1. Slime molds bear little resemblance to other members of the kingdom Fungi

2. During part of their life cycle, slime molds lack cell walls and consist of a multinucleate mass of cytoplasm called a **plasmodium**

   a. Plasmodia often appear white, but can be blue, orange, yellow, or black

   b. They flow or creep along substrates much like the protozoan amoeba

   c. Plasmodia contain many (sometimes several thousand) diploid nuclei

3. Under stressful environmental conditions (drought or starvation), plasmodia are converted into a stationary structure from which sporangia begin to form

4. Sporangia, tiny spheres held above the substrate on small stalks, produce spores
5. Meiosis occurs in the individual spores, forming four nuclei, three of which disintegrate; the remaining haploid spore is released and later germinates
6. The germination of the haploid spore produces an amoeba-like cell known as a *myxamoeba* or a flagellated cell called a swarm cell
7. After feeding on bacteria and other small organisms, the myxamoebae or swarm cells fuse to form a diploid zygote, from which a new plasmodium develops, thereby completing the cycle

IV. Importance

A. General information
1. Naturally occurring fungi are important in the decomposition of dead organisms and also can cause disease
2. Cultivated fungi are used as food and as sources of antibiotics and other valuable substances

B. Edible varieties
1. Edible fungi include truffles and morels (sac fungi), *Agaricus bisporus* or cultivated mushrooms (club fungi), and shiitake mushrooms from China and Japan (also club fungi)
2. The sac fungus yeast (*Saccharomyces cerevisiae*) is used in baking and fermentation processes

C. Decomposer varieties
1. In nature, fungi serve as decomposers, breaking down dead plant and animal matter into their component organic materials
2. As decomposers, fungi play an important role in sustaining ecosystems
   a. They permit recycling of nutrients bound up in the tissues of organisms
   b. These nutrients would otherwise be unavailable to sustain new growth

D. Intoxicating, poisonous, and hallucinogenic varieties
1. The *Conocybe* and *Psilocybe* mushrooms have intoxicating and hallucinogenic properties; they were used by the ancient Mayans for religious ceremonies and are still used by the native peoples of Mexico and Central America
2. *Amanita phalloides* and *Amanita verna* (collectively known as the destroying angel, death angel, or death cup) are extremely poisonous; both species appear similar to edible varieties, but can be lethal even if ingested in small quantities

Study Activities

1. Define the terms hypha, mycelium, coenocytic, dikaryotic, and spore.
2. Discuss the importance of mycorrhizae to plants.
3. Discuss the importance of fungi to humans.
4. Describe the life cycles of zygomycetes, ascomycetes, basidiomycetes, and slime molds.

11

Kingdom Plantae: Nonvascular Plants

Objectives

After studying this chapter, the reader should be able to:
- Describe the most important characteristics of members of the Plant kingdom.
- Explain the concept of alternation of generations.
- Describe the ecologic importance of bryophytes.
- Describe the life cycles of liverworts, hornworts, and mosses.

I. Introduction to the Plant Kingdom

A. General information
1. All member of the plant kingdom are multicellular organisms composed of eukaryotic cells
2. All plants contain chlorophylls a and b, and many possess carotenoid as well
3. Plant cells use starch as the storage carbohydrate and cellulose as the structural polysaccharide in cell walls
4. Plant reproduction is primarily sexual, although asexual reproduction occurs
5. Evolutionary trends in the Plant kingdom are related to the modifications or adaptations necessary for living in terrestrial environments; for example, because dehydration is a possibility in such environments, plants have developed methods of controlling water loss

B. Alternation of generations
1. *Alternation of generations* is the typical sexual life cycle of a plant in which a sporophyte generation alternates with a gametophyte generation
2. The *sporophyte generation* is the spore-producing diploid generation of the life cycle
   a. Spores are produced by meiosis within the sporophyte
   b. These spores undergo cell division to form a multicellular gametophyte
3. The *gametophyte generation* is the gamete-forming haploid generation of the life cycle
   a. The gametes (egg and sperm) are produced by mitosis and contained within a sterile jacket of cells to prevent desiccation (drying out)
   b. The haploid gametes unite through fertilization to produce a diploid zygote, or sporophyte
4. Plant divisions vary as to which generation (sporophyte or gametophyte) predominates

II. Nonvascular Plants (Bryophytes)

A. General information
1. Nonvascular plants lack the vascular and transport tissue (xylem and phloem) found in other plants
2. These plants are members of the division Bryophyta and are found in cool, shaded, extremely moist areas; however, many **bryophytes** can withstand long periods of desiccation
3. Bryophytes are ecologically important in forming bogs (peat moss) and in developing ecologic communities, a process known as **succession**
4. Many bryophytes have associated mycorrhizal fungi
5. They have flagellated, swimming sperm and require water for successful fertilization
6. They spend most of the life cycle as gametophytes; the sporophyte remains attached to and lives off the gametophyte
7. The three classes of bryophytes are liverworts, hornworts, and mosses

B. Class Hepaticae (liverworts)
1. Liverworts consist of a thin, flattened structure called a **thallus**
   a. The thallus develops directly from a spore
   b. Its smooth upper surface has various markings and pores; the underside bears many tiny, single-cell, rootlike structures known as *rhizoids*, which anchor the thallus to the substrate
2. Representative liverworts are *Marchantia* (common liverwort) and *Porella* (leafy liverwort)
3. The life cycle of the liverwort is typified by *Marchantia*
   a. Asexual reproduction occurs by fragmentation and the production of **gemmae**, clusters of lens-shaped cells that detach from the parent
      (1) Gemmae are produced in specialized cuplike structures formed on the upper surface of the thallus
      (2) The gemmae are dispersed by rainfall, and each gemma is capable of growing into a new thallus
   b. Sexual reproduction begins with the gametophyte generation
      (1) The male and female thalli of the gametophyte generation produce male and female gametangia
      (2) The *gametangia* are upright, umbrella-like structures that extend above the thallus and produce gametes within specialized structures known as **antheridia** and *archegonia*
         (a) The numerous antheridia within the upper surface of the male gametangium produce many sperm
         (b) The archegonia on the under surface of the female gametangium contains a single egg
      (3) When the flagellated sperm are released during a rainstorm, they swim to the archegonia to fertilize the egg
      (4) Fertilization produces a zygote that develops into an embryo (immature sporophyte)
      (5) The embryo remains attached to the archegonium as it grows into a mature sporophyte, which consists of three parts: foot (point of attachment to archegonium); seta (short thick stalk); and capsule or sporangium (structure in which meiosis occurs to form haploid spores)

Life Cycle of Moss

The gametophyte generation comprises a protonema and a leafy gametophyte. The sporophyte generation grows from the zygote and remains nutritionally dependent on the gametophyte.

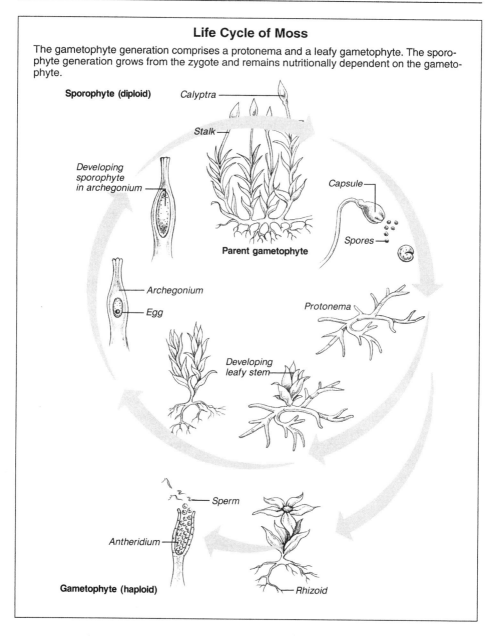

Sporophyte (diploid)   *Calyptra*

Stalk

Developing sporophyte in archegonium

Capsule

Spores

Parent gametophyte

Archegonium

Egg

Protonema

Developing leafy stem

Sperm

Antheridium

Gametophyte (haploid)   *Rhizoid*

(6) At maturity, the spores are released from the capsule and form a new thallus when they reach a suitable habitat

C. Class Anthocerotae (hornworts)

1. Hornworts resemble liverworts, but during the sporophyte generation, they form hornlike structures that give the class its name
2. The hornwort thalli often produce a mucilage secretion in which nitrogen-fixing cyanobacteria grow

a. This relationship provides the hornwort with a source of nitrogen

b. The presence of cyanobacteria may also provide some coloration

D. Class Musci (mosses)

1. Mosses consist of leaflike structures that contain photosynthetic material but lack a midrib with vascular tissue and therefore are not considered true leaves

2. These low-growing green plants typically are found in moist, shady habitats

3. Asexual reproduction occurs by fragmentation

4. Sexual reproduction begins with the germination of a haploid spore (see *Life Cycle of Moss*, page 89)

   a. The spore forms an algae-like, green filamentous structure called a ***protonema***

   b. The protonema eventually grows to become a leafy gametophyte, which produces male and female gametangia in its sperm-producing antheridia and egg-producing archegonia

   c. The zygote that results from fertilization develops into an embryo while still within the archegonium

   d. The embryo grows to form a sporophyte

   e. Haploid spores produced by meiosis within the sporophyte capsule are released when the operculum (cap) falls off

   f. They are dispersed by wind and eventually germinate to complete the life cycle

E. Importance

1. Bryophytes may have been the first plants to grow on barren or bare surfaces

2. They are a chief source of food in the arctic circle, where they are eaten by reindeer, caribou, and other animals

3. Mosses retain moisture and can potentially reduce flooding and erosion in natural ecosystems

4. They also contribute to the humus content of soil; for example, peat moss is used as a soil conditioner to promote moisture retention and increase the nutrient status of soils

---

Study Activities

1. Name the principal characteristics of members of the Plant kingdom.

2. Illustrate the concept of alternation of generations and explain how the two generations are connected.

3. Explain why bryophytes are small and confined to wet areas.

4. Describe the life cycles of liverworts, hornworts, and mosses.

5. List the economic uses of bryophytes.

12

Nonseed Vascular Plants: Ferns and Their Relatives

Objectives

After studying this chapter, the reader should be able to:
- Explain the difference between nonseed and seed vascular plants.
- Name the major divisions of nonseed vascular plants.
- Describe the life cycles of whisk ferns, club mosses, horsetails, and ferns.
- List the uses of nonseed vascular plants.

I. Vascular Plants

A. General information
1. Vascular plants contain conducting tissue consisting of xylem and phloem
   a. Vascular tissue allows plants to inhabit drier habitats
   b. The ability to conduct water through the plant body permits some vascular plants to grow extremely large (such as giant redwood trees)
2. Unlike the bryophytes, vascular plants possess true roots, stems, or leaves
3. Many vascular plants can simultaneously control water loss and permit gas exchange because of a waxy outer covering (cuticle) and small porelike openings on the leaf (stomata)
4. The diploid sporophyte generation is the predominant generation in vascular plants; the gametophyte is small, short-lived, and often inconspicuous

B. Taxonomic classification
1. Vascular plants are classified according to the type of **propagule** produced
2. Nonseed vascular plants reproduce using spores; this group includes ferns and their relatives
3. Seed vascular plants reproduce using seeds; this group includes gymnosperms and angiosperms

II. Nonseed Vascular Plants

A. General information
1. Nonseed vascular plants are categorized according to the type of spores that they produce

2. **Homosporous** plants produce only one kind of spore in a single sporangium
   a. The spore undergoes cell division to produce bisexual gametophytes that have both antheridia and archegonia
   b. The gametophytes develop outside the spore wall
   c. Homospory is characteristic of whisk ferns, horsetails, some club mosses, and most ferns
3. **Heterosporous** plants produce two types of spores in two different sporangia
   a. Heterospory is characteristic of some club mosses, a few ferns, and all seed plants
   b. The two types of spores, called **microspores** and **megaspores**, are much larger than those produced by homosporous vascular plants
      (1) Microspores, which may or may not be smaller than megaspores, give rise to male gametophytes
      (2) Megaspores give rise to female gametophytes
   c. The free-living, nutritionally independent gametophytes develop within the spore wall
   d. The gametophytes require water for fertilization so that the flagellated sperm can reach the egg

B. Taxonomic groups

1. The four divisions of nonseed vascular plants are Psilophyta, Lycophyta, Sphenophyta, and Pterophyta
2. With the exception of Pterophyta (ferns), most of these plants are uncommon, unfamiliar species

III. Division Psilophyta (Whisk Ferns)

A. General information

1. Psilophyta are tropical and subtropical plants that consist of only one existent family, Psilotaceae, with just two genera: *Psilotum* and *Tmesipteris*
2. In the United States, whisk ferns grow naturally in Florida, Louisiana, Texas, Arizona, and Hawaii
3. They have no economic importance, although florists occasionally use them in arrangements

B. Structure

1. Whisk ferns do not have true leaves or roots; photosynthesis occurs in the outer epidermal cells of the stem
2. The dichotomously forking stems develop from rhizomes that grow horizontally just under the soil surface
3. The stem and branches contain a central core of vascular tissue, with phloem surrounding a star-shaped core of xylem
4. Surrounding the stele is an area of cortex comprised of parenchyma and sclerenchyma tissue
5. These homosporous plants produce spores within sporangia, which are clustered on the stem and branches

C. Reproduction

1. The life cycle of the whisk fern reflects the alternation of generations found in all plants
2. Spores released from the sporangia of the sporophyte germinate and grow to form a gametophyte within the soil, on the bark of trees, or on other surfaces
   a. The colorless, saprophytic gametophyte lacks chlorophyll and obtains nutrients by association with fungi
   b. The gametophyte is approximately 2 mm in diameter and 6 mm long
3. Archegonia and antheridia are produced randomly on the surface of the gametophyte
4. Fertilization requires water so that the flagellated sperm can swim to the archegonia
5. The fertilized egg, now a diploid zygote, develops into an embryo within the archegonium
6. As the embryo continues to grow, it forms a foot and a shoot apex
   a. The foot permits the growing sporophyte to obtain nourishment from the gametophyte
   b. The shoot apex undergoes repeated cell division to form the mature sporophyte
7. The sporophyte ultimately detaches from its foot and lives independently of the gametophyte
   a. The sporophyte consists of an underground rhizome, which performs the same functions as a root, and a series of upright aerial branches that arise from the rhizome
   b. As the rhizome grows, it becomes infected with mycorrhizae, which help the rhizome anchor the plant and absorb water and nutrients from the soil
8. Sporangia develop in clusters on the upright, dichotomously forked branches of the sporophyte
9. Spores are released from the sporangia, and the cycle is repeated

IV. Division Lycophyta (Club Mosses)

A. General information

1. The principal club mosses are *Lycopodium, Selaginella,* and *Isoetes*
   a. *Lycopodium* (ground pine) commonly grow on forest floors in temperate climates
   b. *Selaginella* (spike mosses) grow in wet areas throughout the world, but are common in the tropics and are occasionally a weed in greenhouses
   c. *Isoetes* (quillworts) grow partially submerged in marshy areas
2. Treelike ancestral varieties of lycophytes were dominant plants in the forests of the Carboniferous period some 286 to 360 million years ago; their fossilized remains helped produce fossil fuels (coal and petroleum) used today
3. Historically, club moss spores were used to produce flash powders and explosives; various parts of the plant were also used in folk medicine and remedies
4. Club mosses are currently used to create Christmas decorations and wreaths; they are so popular with crafts people that several species are now endangered from over-collection
5. Many species are an important food source for wildlife

Life Cycle of the Fern

The sporophyte generation is the most recognizable form of a fern. The gametophyte generation consists of a small, heart-shaped structure found littered on the forest floor. The mature plant (sporophyte) produces fronds with sporangia attached. The spores produced in the sporangia germinate to form the gametophyte.

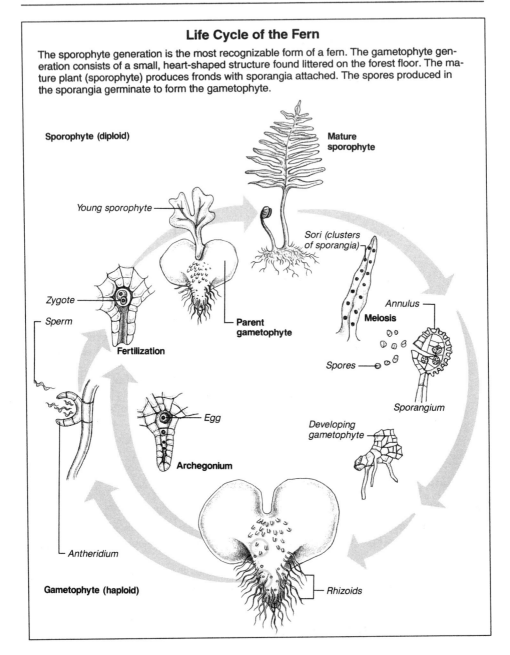

Sporophyte (diploid)

Mature sporophyte

Young sporophyte

Sori (clusters of sporangia)

Zygote

Sperm

Annulus

Meiosis

Parent gametophyte

Fertilization

Spores

Sporangium

Egg

Developing gametophyte

Archegonium

Antheridium

Gametophyte (haploid)

Rhizoids

B. Structure

1. The upright or occasionally prostrate stems develop from branched rhizomes; the stems are covered with small leaves called **microphylls**
2. Adventitious roots develop along the rhizome, allowing segments of the rhizome to become independent plants
3. Sporangia typically are located in terminal clusters called **strobili** or on specialized leaves called **sporophylls**

a. Ground pine sporophytes are generally less than 30 cm tall

b. Quillwort sporophytes are generally less than 10 cm tall

C. Reproduction

1. The life cycle of the club moss begins with the production of spores within the sporangia of the sporophyte

2. The spores are dispersed and germinate to form independent gametophytes, which typically grow on the soil surface

3. When mature, the gametophytes spawn both antheridia and archegonia, which produce sperm and eggs

4. Water is required for the flagellated sperm to swim from the antheridia to the archegonia so that fertilization can occur

5. Fertilization results in a zygote with foot, stem, and leaves

V. Division Sphenophyta (Horsetails and Scouring Rushes)

A. General information

1. Historically, horsetails were used by native Americans for medicinal purposes, food, and scouring pads (for cleaning cooking utensils)

2. *Equisetum*, which is typically found along streams, water bodies, and wetlands, is the most common genus

B. Structure

1. *Equisetum* are tall, reedlike plants that resemble a horse's tail

2. Stems develop from horizontal rhizomes, which can be quite extensive and form large horsetail colonies

a. The distinctly ribbed, jointed stems are the site of photosynthesis

b. The stems have scalelike leaves that occur in whorls at regular intervals

c. The stems of the sporophyte contain silica deposits on the inner walls of the epidermal cells

C. Reproduction

1. Asexual reproduction occurs by fragmentation of the stem or rhizome; if these structures are broken up by a disturbance, such as a storm or foraging animals, the fragments can grow into new sporophytes

2. Sexual reproduction begins with the formation of strobili at the stem tips

a. The small, conelike strobili bear the sporangia in which spores are produced

b. After dispersal, the spores produce small green gametophytes that generate archegonia and antheridia

c. Because the sperm are flagellated, they need water to swim to the archegonia

d. When the egg is fertilized, a zygote forms and develops into the new sporophyte generation

VI. Division Pterophyta (Ferns)

A. General information
    1. Historically, ferns have been used as food and medicine and were often an ingredient in folk remedies
    2. Today ferns are commonly used as houseplants and ornamental plants
    3. In natural ecosystems, they are an important food source for wildlife

B. Structure
    1. Ferns vary in size from small floating forms less than 1 cm in diameter to large, tropical, treelike ferns up to 25 meters tall
    2. These plants possess large, highly divided, feathery leaves called **fronds**
       a. Fronds first appear as small, tightly coiled structures called *fiddleheads*, which uncurl to form mature fronds
       b. There are two types of fronds
          (1) *Vegetative fronds* are involved only in photosynthesis
          (2) *Reproductive fronds* have sporangia for the production of spores

C. Reproduction
    1. Spore production begins in the sporangia, which are often found in clusters (called *sori*) on the underside of fronds (see *Life Cycle of the Fern*, page 94)
       a. In many ferns, the sori are protected by flaps of covering tissue called an **indusium**; as the sporangia mature, the indusium shrivels to expose the sporangia beneath it
       b. Each sporangium has a conspicuous row of thick-walled cells along one edge, which are known as the **annulus**
       c. The annulus catapults spores out of the sporangium using a snapping action that is influenced by moisture changes within the cells
       d. The individual sporangia contain diploid cells (parental cells) that undergo meiosis to produce haploid spores
    2. Released spores germinate and form a heart-shaped **prothallus**
    3. The prothallus is a multicellular, independent, photosynthetic gametophyte
    4. Archegonia develop on the under surface of the prothallus near a notch in the heart shape; antheridia develop near the apex
    5. After fertilization, the zygote formed in the archegonium continues to grow into an independent sporophyte, which is the recognizable fern

---

Study Activities

1. Compare and contrast vascular and nonvascular plants.
2. Compare and contrast seed and nonseed plants.
3. Explain why nonseed vascular plants are better adapted to terrestrial environments than bryophytes.
4. Describe the life cycle of a plant from each division of the nonseed vascular plants.
5. Explain why nonseed vascular plants depend on water for reproduction.
6. Name the human uses of the plants discussed in this chapter.

13

Seed Vascular Plants: Gymnosperms

Objectives

After studying this chapter, the reader should be able to:
- Define gymnosperm.
- Differentiate between gymnosperms and angiosperms.
- List and describe the major divisions of gymnosperms.
- Describe the life cycle of a pine tree.
- Discuss the human and ecologic importance of gymnosperms.

I. Seed Vascular Plants

A. General information

1. Seed plants are so named because the seed is the chief reproductive propagule
   a. A seed consists of an embryo that is packaged with stored food and surrounded by a protective seed coat; the stored food is used for energy during germination and early seedling development
   b. The seed permits the plant to survive harsh environmental conditions (cold or lack of moisture) during a period of seed dormancy
2. All seed vascular plants have conducting tissue consisting of xylem and phloem
3. All seed-bearing plants exhibit alternation of generations
   a. The gametophyte generation in these plants is extremely small and develops within the tissues of the sporophyte
      (1) The male gametophyte is the pollen grain
      (2) The female gametophyte is the embryo sac
   b. All seed plants are heterosporous (produce two kinds of spores); the megaspore gives rise to the female gametophyte and the microspore gives rise to the male gametophyte
   c. Seed plants generally do not require water for fertilization because, in most varieties, the sperm are not flagellated and do not swim to the archegonia

B. Taxonomic classification

1. Seed plants are classified as gymnosperms or angiosperms
   a. **Gymnosperms** produce seeds in structures open to the environment
   b. **Angiosperms** (flowering plants) produce seeds in specialized reproductive structures called flowers
   c. Gymnosperms are primarily trees and shrubs; angiosperms encompass a diverse variety of life forms

2. The four divisions of gymnosperms are Coniferophyta, Cycadophyta, Ginkgophyta, and Gnetophyta

II. Division Coniferophyta (Conifers)

A. General information
1. Conifers are the largest group of gymnosperms
2. The more than 575 species of conifers include pines (*Pinus*), firs (*Abies*), spruces
   (*Picea*), hemlocks (*Tsuga*), cypresses (*Cupressus*), Douglas firs (*Pseudot-
   suga*), and junipers (*Juniperus*)

B. Pines
1. Most pines are found in the northern hemisphere, although they have been extensively planted in the southern hemisphere
2. Needles are a distinctive characteristic of the pines
   a. They usually appear in clusters known as **fascicles**
   b. They have adapted to winter conditions, when the ground water is frozen
      and the needles are exposed to drying winds
      (1) The needles have a thick epidermis with a heavy cuticle
      (2) Several layers of thickened cells just below the epidermis (called the
          hypodermis) further reduce the potential for water loss
      (3) Stomata are located in pits to shield them from the wind's drying action
      (4) Veins are surrounded by endodermis, which is another layer of large
          cells that prevents excessive water loss
      (5) Mesophyll cells are packed tightly together; they do not have the obvi-
          ous air spaces seen in the spongy mesophyll cells of deciduous leaves
      (6) The presence of resin canals, which are produced in response to in-
          jury, may prevent insect and fungal damage
   c. The fascicles drop off two to five years after maturation; they are lost a few at
      a time rather than all at once as with the leaves of deciduous trees
3. The xylem tissue of conifers contains tracheids as the conducting cells instead of
   the vessels found in angiosperms
   a. Because the secondary xylem of conifers lacks thick-walled vessels and fi-
      bers, conifer woods are labeled soft woods
   b. The resin canals in the secondary xylem serve to inhibit fungal infections
      and prevent damage by plant-eating animals (**herbivores**)
4. The thick bark of conifers contains large amounts of secondary phloem; the
   phloem contains albuminous cells that perform the same function as the companion cells of angiosperms
5. Pines almost always have mycorrhizae attached to their roots
6. As a heterosporous plant, the pine begins its reproductive life cycle with the production of two types of spores—male microspores and female megaspores
   (see *Life Cycle of the Pine*)
   a. Microspores are produced in a process called **microsporogenesis**
      (1) Male cones, which are typically 1 to 4 cm long, usually are found in
          clusters of 50 or more on tips of the lower branches
      (2) Male cones last for only a few weeks, during which time each diploid
          microspore parent cell, or *microsporocyte*, undergoes meiosis to form
          four haploid microspores

Life Cycle of the Pine

The typical life cycle of a pine takes nearly two years to complete. The female gametophyte generation comprises the egg and its associated structures; the male gametophyte consists of the germinated pollen grain. The tree is considered the sporophyte.

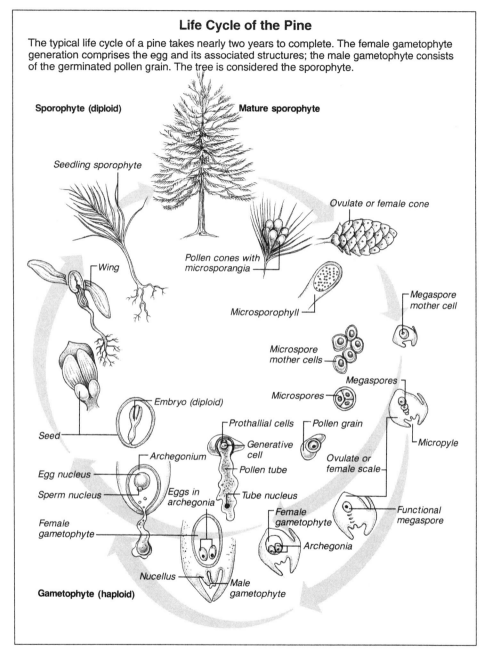

Sporophyte (diploid)

Mature sporophyte

Seedling sporophyte

Ovulate or female cone

Wing

Pollen cones with microsporangia

Microsporophyll

Megaspore mother cell

Microspore mother cells

Megaspores

Microspores

Embryo (diploid)

Seed

Prothallial cells

Pollen grain

Generative cell

Archegonium

Pollen tube

Ovulate or female scale

Egg nucleus

Sperm nucleus

Eggs in archegonia

Tube nucleus

Micropyle

Female gametophyte

Female gametophyte

Functional megaspore

Archegonia

Nucellus

Male gametophyte

Gametophyte (haploid)

(3) Each haploid microspore undergoes mitosis and matures into a pollen grain, which consists of four to five cells and a pair of external air sacs or wings

(4) The pollen grain contains a tube cell (which forms the ***pollen tube***); a generative cell (which, after dividing, produces sperm nuclei); and two

to three prothallial cells (which are the remnant vegetative cells of the gametophyte generation)

(5) Vast numbers of pollen grains are shed from the cones and dispersed by the wind to pollinate the female cones

(6) To aid wind pollination, pollen grains have two wings that develop from the microspore cell wall

(7) Once the pollen grains are released, the male cones dry up and drop from the tree

b. Megaspores are produced in a process called *megasporogenesis*

(1) Female cones are larger than male cones and develop in the spring

   (a) These woody cones persist for long periods of time

   (b) They are generally located on the upper branches

(2) The female cone consists of a number of modified leaflike organs called *sporophylls* that bear sporangia and serve as shelves on which the seeds develop

(3) Two **ovules** are located on each sporophyll

   (a) Each ovule has an opening, called a **micropyle**, through which the pollen tube enters, and a **megasporangium** (also called the **nucellus**) surrounded by two **integuments** (outer tissue layers)

   (b) Each megasporangium contains a **megasporocyte**, or megaspore parent cell

(4) The megasporocyte produced within the ovule undergoes meiosis to form four haploid megaspores; three disintegrate and one develops into the female gametophyte

(5) The female gametophyte grows slowly, taking approximately 13 months to reach maturity

(6) Once mature, the female gametophyte consists of several thousand cells and two to six archegonia, each containing a single egg

(7) Typically all eggs in the archegonia are fertilized and begin to develop into embryos (a phenomenon called *polyembryony*, or multiple embryo production); only one embryo develops fully and the others spontaneously abort

7. Fertilization and seed development occur within the female pine cone

a. In the first spring of the reproductive cycle, the immature cone scales spread apart; pollen grains (carried by air currents) sift down between the sporophylls and are caught in sticky secretions near the micropyle

b. The pollen grain germinates within the micropyle to produce a pollen tube

c. As the pollen tube grows toward the archegonium, the generative cell of the pollen grain divides, producing a sterile cell and a spermatogenous cell

(1) The spermatogenous cell further divides to form two sperm; the sterile cell, whose function is unknown, disintegrates

(2) The mature male gametophyte consists of the pollen grain and the two sperm

d. About 15 months after pollination, the pollen tube reaches the archegonium and releases its contents

e. One sperm unites with the egg and the other disintegrates

f. The zygote formed from the union develops into an embryo

g. The developing embryo is surrounded by integument, which consists of remnant sporophyte tissue and forms a seed coat

       h. The mature embryo consists of an epicotyl, hypocotyl, radicle, and multiple cotyledons

  8. In pines, the seed matures approximately 12 months after fertilization, and the complete life cycle takes about two years

       a. Male and female cones are formed during the summer of the first year

       b. In the spring of the following year, microspores and megaspores are produced by meiosis

       c. Pollination occurs during the early summer, and fertilization takes place in the late spring of the second year

       d. By the summer of the second year, the embryo develops

       e. Seeds are produced and dispersed in the fall of the second year

  9. Seeds can remain dormant for many years, and some may be embedded in the mature cones for six years or more

  10. Pine seeds are shed from the cones during the autumn of the second year

       a. When mature, the cone scales of most pine species separate, and the winged seeds are carried through the air by wind currents

       b. In species whose seeds do not have wings, the seeds may be dropped by birds during flight or when attempting to eat

       c. In some species, such as the lodgepole, jack, and knobcone pines, the cones open only when exposed to heat (a fire, for example), which melts the resin that holds the cone scale closed

  11. Other conifers lack the needle clusters of pines and may have slightly different reproductive cycles

       a. The yews (family Taxaceae) do not produce woody female cones; instead, the ovule is at least partially surrounded by a fleshy, cuplike covering

       b. The Norfolk Island pine (*Araucaria excelsa*) lacks the needle clusters of other pines and bears its needles singly along stems and branches; it is native to the southern hemisphere and is a common houseplant

III. Other Gymnosperms

A. General information

  1. In addition to the conifers, a number of tropical and subtropical plant species are classified as gymnosperms

  2. Other gymnosperms that have economic value or unique characteristics include the Cycadophyta, Ginkgophyta, and Gnetophyta

B. Division Cycadophyta (cycads)

  1. Cycads are slow-growing tropical and subtropical plants that consist of unbranched trunks with a crown of pinnately divided (palmlike) leaves

  2. Cycads are ***dioecious*** (pollen and seeds are produced on different plants)

  3. Their sperm are atypical because they have numerous flagella

C. Division Ginkgophyta (ginkgoes)

  1. The only living species in this division is *Ginkgo biloba*, or the maidenhair tree

  2. This dioecious tree has broad, notched, fan-shaped leaves with no prominent midrib; it is commonly planted along sidewalks in the United States

  3. Its seeds resemble small apricots or plums, but produce a rank odor because of the butyric acid in the seeds' fleshy coat

4. The sperm are delivered to the egg through a pollen tube

D. Division Gnetophyta
1. This unique group of heterogeneous gymnosperms shares some characteristics with angiosperms, the most important of which are the presence of vessels (in addition to tracheids) in xylem tissue and the similarity of their strobili to angiosperm inflorescences
2. The three living genera of Gnetophyta are *Gnetum*, *Ephedra*, and *Welwitschia*
   a. *Gnetum* is a woody vine with broad leaves; it is native to the tropics
   b. *Ephedra* is a shrublike plant with jointed stems and small leaves; it is usually found in desert or semidesert habitats
   c. *Welwitschia* is a rare and exotic plant found only in the coastal regions of Angola and southwest Africa

IV. Human and Ecologic Importance

A. General information
1. The most exploited group of gymnosperms are the conifers, which have been used for food, paper, lumber, and ornamental purposes
2. Ecologically, spruces and firs are the predominant species of the northern coniferous forests (called *taiga*)

B. Conifers
1. Historically, conifers were used by native Americans, who ate the bark raw or dried and ground it into a flour, and by early New England settlers, who made a tea high in vitamin C from the needles
2. Resins extracted from conifers are used to produce turpentine, menthol, floor waxes, printing ink, paper coatings, varnishes, and perfumes
3. About 40 species of firs (particularly the Douglas fir) are widely used in the construction, plastic, and paper industries
4. Spruces, firs, eastern white cedars (arborvitae), and pines are used as Christmas trees and ornamentals
5. Spruces are an important source of pulpwood for newsprint
6. Cedars are harvested to make pencils

C. Other gymnosperms
1. Ginkgo seeds are commonly eaten in the Orient
2. Many other gymnosperms, such as yews, are used as ornamental plants

---

Study Activities

1. List the major characteristics of gymnosperms.
2. Describe all four divisions of gymnosperms and give an example of each.
3. Explain how the pine needle can survive dry conditions.
4. Describe the life cycle of a pine tree.
5. List at least four uses of gymnosperms.

14

Flowering Plants: Angiosperms

Objectives

After studying this chapter, the reader should be able to:
• Describe the characteristics of angiosperms.
• Describe the life cycle of angiosperms.
• List the various methods of pollination.
• Explain double fertilization.
• Explain apomixis and the formation of parthenocarpic fruits.
• Distinguish between monocots and dicots.
• Discuss the major evolutionary trends involving angiosperms.

I. Flowering Plants

A. General information
    1. The flowering plants, or angiosperms, belong to the division Anthophyta—it is the largest division of the plant kingdom, containing more than 240,000 species
        a. The flower is a distinguishing characteristic of this group
        b. Angiosperm seeds are contained within a vessel-like, female structure called the *carpel* (Note that some texts use the term *carpel* in place of *pistil*)
        c. The angiosperm ovary develops into a fruit
    2. Angiosperms have small, brief gametophyte generations

B. Female gametophyte
    1. In megasporogenesis, megaspores are formed within the megasporangium (nucellus) and ultimately develop into the female gametophyte (see *Life Cycle of an Angiosperm*, page 104)
    2. The ovule is composed entirely of nucellus at the beginning of the reproductive cycle, but it soon begins to differentiate
        a. During ovule development, one or two enveloping layers called integuments are formed
        b. Each integument has a small opening (micropyle) at one end through which the pollen tube enters
    3. While the flower is still in bud, a diploid megaspore parent cell forms in the ovule and subsequently undergoes meiosis to produce four haploid megaspores, of which three disintegrate
    4. The remaining megaspore undergoes a series of nuclear divisions without subsequent cytoplasmic division, resulting in a cell with eight haploid nuclei

Life Cycle of an Angiosperm

The gametophyte generation of angiosperms is small and inconspicuous. The male gameto-phyte is the germinated pollen grain, and the female gametophyte is the mature embryo sac. After the zygote is formed, it develops into a seed; the ovary wall may swell, thereby forming a fruit to aid in seed dispersal.

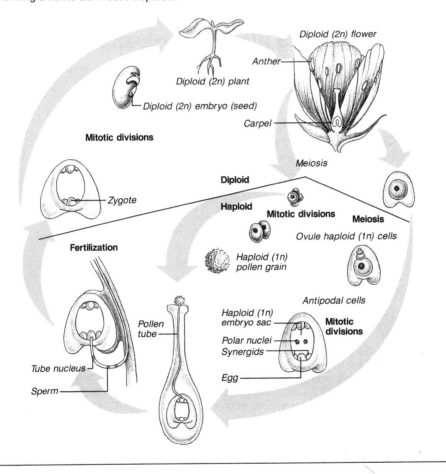

5. This eight-nuclei stage comprises the female gametophyte, or embryo sac
   a. It consists of two polar nuclei located in the middle of the cell and a single egg cell with two "helper cells" (called *synergids*) at the micropyle end
   b. Three antipodal cells of unknown function are located at the end opposite the egg

C. Male gametophyte
1. The male gametophyte is the germinated pollen grain
2. Its formation begins within the anther, where the microspore parent cells undergo meiosis to form haploid microspores
3. The microspore nucleus divides to produce two nuclei—the tube nucleus and the generative nucleus

    a. The tube nucleus is contained within a pollen **tube cell**, which is responsible
       for germination of the pollen grain and growth of the pollen tube toward the
       micropyle

    b. The generative nucleus is contained within a **generative cell**, which is re-
       sponsible for the formation of sperm when the pollen grain germinates

4. The thick cell wall that develops around the mature pollen grain is resistant to
   moisture loss; it protects and preserves the tube and generative nuclei during
   pollination

D. Pollination

1. **Pollination** is the transfer of pollen from the stamen to the stigma of a flower, and
   often occurs from plant to plant

2. Plants use a variety of mechanisms to transfer pollen, including physical forces,
   such as wind or water currents, and reliance on specific organisms, such as insects, birds, and mammals

3. *Wind-pollinated flowers* are typically small and inconspicuous and have well-
   exposed anthers and stigmas

    a. They tend to produce large amounts of pollen because much of it is lost dur-
       ing transfer owing to the unpredictability of wind direction and air currents

    b. They generally do not produce odors for attraction or provide rewards for pol-
       linators

    c. Examples include walnut, grasses, and corn

4. *Insect-pollinated flowers* must have a mechanism to attract and reward the pollinators

    a. Bee-pollinated flowers produce a sweet nectar and pleasant fragrances and
       are usually brightly colored (predominately blue or yellow, rarely red)

       (1) These flowers often have lines or other distinctive markings visible in
          the ultraviolet range that may lure the bees to the nectar

       (2) Bees often consume nectar and pollen as rewards for pollination

       (3) Examples include larkspur, snapdragon, violet, rosemary, foxglove, and
          clover

    b. Beetle-pollinated flowers have strong, yeasty, spicy, or fruity odors; such
       odors are important because beetles have poor eyesight but a keen sense
       of smell

       (1) These flowers are white or dull-colored and usually produce no nectar

       (2) They often provide pollen as a food source for the beetle or possess
          special food glands; occasionally the beetles chew flower parts

       (3) Examples include magnolia and dogwood

    c. Fly-pollinated flowers have foul odors to attract the insects

       (1) These flowers are usually dull red or brown and provide nectar as a re-
          ward for pollination

       (2) They tend to remain wide open to allow easy access by flies

       (3) Examples include carrion flower, certain cacti, skunk cabbage, and
          some lilies

    d. Moth-pollinated flowers have strong, sweet fragrances that are released at
       night when moths are most active

       (1) These flowers are pale white or yellow and provide nectar as a reward

       (2) Examples include tobacco, Easter lily, some cacti, evening primrose,
          and amaryllis

e. Butterfly-pollinated flowers have sweet fragrances and are often bright red, orange, or yellow
  (1) These flowers produce fragrances during the day, when butterflies are active, and provide nectar as a reward
  (2) Examples include the butterfly weed and daisy
5. *Animal-pollinated flowers* must provide larger rewards because of the greater energy needs of the pollinators
  a. Bat-pollinated flowers are dull colored and produce strong, perfume-like odors
    (1) These flowers open at night when bats are active and are large enough for the bat to insert its head in the flower
    (2) They produce large amounts of nectar; bats also eat pollen and other flower parts
    (3) Examples include organ pipe cactus and numerous tropical species
  b. Bird-pollinated flowers are usually bright red or yellow and produce little or no odor
    (1) The large or numerous, usually tubular-shaped flowers attract hummingbirds or sunbirds by providing nectar as a reward
    (2) Examples include red columbine, fuchsia, hibiscus, passion flower, eucalyptus, and poinsettia
6. Pollen can be transferred within individual flowers of the same plant or between flowers of plants separated by great distances
  a. **Self-pollination** is the transfer of pollen from stamen to stigma within the same flower or between the flowers of the same plant; it reduces the chances of genetic variation
  b. **Cross-pollination** is the transfer of pollen from stamen to stigma of different, genetically distinct flowers; it promotes genetic variation within the population, which helps plants adapt to changes in the environment and ensures the survival of the species

E. Fertilization and seed development
1. Pollen grains transferred during pollination adhere to the stigma of the pistil, where they germinate
2. Germination produces a generative cell that divides to form two sperm cells and the tube nucleus
3. The pollen tube grows toward the ovary by digesting its way through the style of the pistil
4. When the pollen tube reaches the embryo sac, it triggers a unique set of events called **double fertilization**—the simultaneous fusion of the two sperm cells in the pollen tube with the cells of the embryo sac
  a. In double fertilization, the tube nucleus disappears and the two sperm move into the embryo sac
  b. One sperm fuses with the egg and forms a zygote
  c. The other sperm fuses with the two polar nuclei of the embryo sac to form a triploid endosperm nucleus, which eventually divides to form the endosperm
  d. In some angiosperms, the endosperm becomes an extensive part of the seed; in others, the endosperm is consumed in the formation of cotyledons and thus disappears by the time the seed is mature

e. The seed is a mature ovule that contains an embryo surrounded by a protective seed coat
  (1) The embryo is an immature sporophyte
  (2) The cotyledon, which stores food, is part of the embryo
  (3) The seed coat, which develops from the integuments, is the outer boundary of the seed
  f. In some plants, double fertilization stimulates the ovary wall to thicken and form a fruit, or pericarp

F. Apomixis

1. **Apomixis** is a form of atypical reproduction in angiosperms in which seeds develop without fertilization; the development of other plant parts, such as flowers and fruits, is normal
2. The embryo in unfertilized angiosperms may develop from a nutritive cell of the ovary
3. Fruits that develop from unfertilized ovaries are called **parthenocarpic**
  a. These fruits are usually seedless
  b. Examples include bananas, navel oranges, some figs, and seedless grapes

II. Monocotyledons and Dicotyledons

A. General information

1. The two classes of the division Anthophyta are monocotyledons and dicotyledons, commonly referred to as monocots and dicots
2. Although the terms *monocot* and *dicot* refer specifically to the number of cotyledons, or food storage organs, within the seeds of angiosperms, several other characteristics separate the two classes (see *Comparison of Monocots and Dicots*, page 108)

B. Monocots

1. Monocots have only one cotyledon in each seed
  a. The single cotyledon enzymatically absorbs food from the endosperm and transports it to the growing embryo
  b. The large cotyledon in the mature seeds of grass species is called the *scutellum*
    (1) The scutellum is attached to one side of the embryo, midway between the plumule at the upper end of the embryo and the radicle at the lower end
    (2) The radicle, or embryonic root, is enclosed in a protective sheath known as a **coleorhiza**
    (3) The plumule (epicotyl and young leaves) is enclosed within a protective sheath called the **coleoptile**
  c. The endosperm persists in the seed and is the primary source of stored food for the embryo and young seedling during the early stages of germination
2. The flower parts (sepals, petals, stamen) usually occur in groups of three or multiples of three
3. The leaves have parallel veins
4. Monocots tend to have fibrous root systems
5. The primary vascular bundles in stems are scattered throughout the cross section

Comparison of Monocots and Dicots

CHARACTERISTIC	MONOCOTS	DICOTS
Embryo	One cotyledon	Two cotyledons
Flower parts	Occur in groups of three	Occur in groups of four or five
Leaves	Parallel veined	Net veined
Roots	Many main roots (fibrous root system)	One main root (taproot system)
Stem anatomy	Scattered vascular bundles	Vascular bundles arranged in rings
Root anatomy	Pith	Xylem in center
Secondary growth	No	Yes

6. The roots contain a central core of pith (parenchyma tissue derived from the procambium)
7. True secondary growth is rare or nonexistent because most monocots lack vascular cambium and cork cambium
8. Examples of monocots include cereal grains (such as corn, wheat, and oats), sugarcane, lilies, daffodils, orchids, bananas, palms, and grasses

C. Dicots
1. Dicots have two cotyledons in each seed
   a. The cotyledons are the food storage organs of the embryo
   b. The endosperm does not remain unchanged in dicots; it is converted to a cotyledon during seed development
2. The flower parts occur in multiples of four or five
3. The leaves have netlike veins
4. Dicots tend to have root systems consisting of a single main root (taproot) with small branches
5. The primary vascular bundles in stems form a ring pattern
6. Most dicot roots, with the exception of those of a few herbaceous dicots, do not have a central core of pith
7. True and often abundant secondary growth occurs in many species because of the presence of vascular and cork cambium
8. Examples of dicots include many annual plants (such as tomatoes, peppers, beans, sunflowers, mustards, and common weeds), flowering herbaceous species, and most flowering trees and shrubs

III. Evolutionary Trends

A. General information
1. Angiosperms are the most modern plants, having evolved from gymnosperm ancestors more than 125 million years ago
2. This diverse group is the dominant plant type on earth

B. Angiosperm evolution

1. As plants evolved, the gametophyte generation became progressively smaller and less conspicuous; it is very prominent in primitive plants, such as bryophytes, less so in ferns, and even less so in angiosperms

2. Carpels (ovule-bearing units within the pistils) allow angiosperms to shelter seeds from dehydration and pests and develop a fruit, which aids in seed dispersal

3. Double fertilization allows angiosperms to use energy reserves efficiently because fruit development (and subsequent energy investment) is triggered only after fertilization occurs

4. Most angiosperms rely on animal pollination, which is considered more reliable than wind pollination

5. The specialization of angiosperm flowers, manifested as a reduction and fusion of parts, is considered by some scientists to be a method of saving energy

Study Activities

1. List the features common to all angiosperms.
2. Describe gametophyte development in angiosperms.
3. Describe the process of fertilization.
4. Describe the different methods of pollination.
5. Explain how different types of pollination are suited to different angiosperms.
6. List the differences between monocots and dicots.
7. Describe the evolutionary trends in angiosperms.

15

Plants and People: Helpful and Harmful Interactions

Objectives

After studying this chapter, the reader should be able to:
- List at least ten plant families and name the economically useful plants in each.
- Explain why the preservation of plant species is important to humans.
- Name the principal cereal grains used throughout the world.
- Identify the plant source of at least six drugs.

I. Useful Monocotyledons

A. General information
1. The most important monocotyledons, or monocots, are the grasses, which supply the major food sources for the world's population
2. Many varieties of monocots have been domesticated and are quite different from their naturally occurring ancestral stocks

B. Poaceae (grass family)
1. Grasses are divided into groups based on their use
   a. *Cereals* and annual grasses are important as grain for humans and as forage for animals
   b. *Sod-forming grasses* have been cultivated for use in lawns and meadows
   c. *Bunch grasses* are commonly found in the natural vegetation of semiarid regions
   d. *Ornamental grasses*, such as pampas grass, have been cultivated as a decorative plant
   e. *Sugar-producing species*, such as sugar cane, grow only in the tropics
   f. *Bamboos* are mostly evergreen varieties that are used for construction in many parts of the Far East and as an ornamental plant in warm areas of the United States
2. Cereals are harvested for their seeds; examples include wheat, rice, corn, barley, rye, and oats
   a. *Wheat* (*Triticum*) is the most widely cultivated plant in the temperate zone
      (1) Wheat has adapted to cool, relatively dry climates and is grown on a large scale in the temperate grassland regions of the world

(2) It can be grown in warmer regions as a winter crop, but does not thrive in tropical areas

(3) The wheat grain, or kernel, is high in nutritional value and contains (by weight) 70% carbohydrate, 8% to 16% protein, 2% fat, and some vitamins

    (a) Modern milling methods remove the germ or embryo (which is high in protein and contains all the fat) and the bran (which is high in protein and vitamins and consists of the ovary wall, seed coat, and outermost layer of the endosperm)

    (b) The milling process results in flour, a white powder consisting mostly of carbohydrate that must be artificially enriched

    (c) The starchy endosperm of flour contains small amounts of protein (gluten and liadin), which causes dough to become sticky and elastic, giving flour its desirable baking characteristics

    (d) Sylvester Graham, a 19th-century Massachusetts preacher, was a proponent of using whole grain flour in baking; the graham cracker is named in his honor

b. *Rice* (*Oryza sativa*) is the principal cereal crop of the humid tropics

(1) Rice requires warm temperatures and abundant moisture; most rice is grown in fields that are flooded during part of the growing season

(2) This grain is the dietary mainstay of 50% of the world's population

(3) It has a low protein content (only 7.5% by weight)

(4) Most rice is eaten as a whole grain, that is, as brown rice in which only the hull is removed

(5) Polishing rice removes the outer layers of the seed and the embryo, leaving only carbohydrate

(6) Although rice can be ground into flour, it not suitable for bread making because the amount of protein it contains is insufficient to hold dough together

c. *Corn* or *maize* (*Zea mays*) is the cereal grain native to the western hemisphere

(1) More than half of the corn produced worldwide is used as animal feed

(2) Corn can be ground into cornmeal for human consumption

(3) Corn oil, which is derived from the embryo, is one of the most important vegetable oils used by humans

(4) Unripe ears of corn, especially those from varieties with a high sugar content (such as sweet corn), are eaten as vegetables

(5) Mexicans prepare corn by soaking the kernels in lime water, squeezing off the hulls, and grinding the remainder into a paste to make tortillas

C. Liliaceae (lily family)

1. Useful members of the lily family include lilies, onions, garlic, asparagus, sarsaparilla, meadow saffron, bowstring hemp, and aloe

2. The roots of some lily varieties, such as squills, are the source of rodent poison and heart-stimulating drugs

3. Colchicine, a drug used to treat gout, is obtained from meadow saffron

4. Bowstring hemps, or sansevierias, are common houseplants in the United Sates and are used in Africa as a source of rope fibers

5. The aloe plant provides juice valuable in treating burns; it is also commonly grown as a houseplant

6. Many lily bulbs are edible but must not be confused with those of daffodils and their relatives, which are poisonous

D. Orchidaceae (orchid family)
1. Orchids are prized as decorative houseplants
2. The vanilla orchid is grown in the tropics as a source of true vanilla flavoring, which is extracted from the fruits

II. Useful Dicotyledons

A. General information
1. Dicotyledons, or dicots, are used as a source of food, spices, beverages, and drugs and for ornamental purposes
2. Some dicots are harvested from natural habitats, while others are cultivated

B. Ranunculaceae (buttercup family)
1. Most members of this family are at least slightly poisonous
2. Many species grow wild, but some (such as columbine, anemone, larkspur, clematis, and goldseal) are cultivated as ornamental plants

C. Lauraceae (laurel family)
1. The laurel family is a source of numerous spices and oils
   a. Cinnamon is obtained from the bark of a laurel tree grown in India and Sri Lanka
   b. Camphor oil is derived by distillation of wood chips from an evergreen tree native to China, Japan, and Taiwan; it is used for cold remedies, inhalants, perfumes, and insecticides
   c. The spicy, aromatic wood of the sassafras tree, which is native to the eastern United States, is used as a flavoring (obtained by distilling bark and wood chips) in toothpaste, chewing gum, mouthwash, and soda
   d. The leaves of the laurel tree are the source of sweet bay, a flavoring used in sauces, soups, and meat dishes
2. Avocados are a variety of laurels and are rich in vitamins and iron

D. Papaveraceae (poppy family)
1. Bloodroot, a wild poppy, was used by native Americans as a facial dye, insect repellent, and cure for ringworm
2. Poppies are a source of morphine and codeine derivatives, which are used as pain killers and cough suppressants
   a. Opium is obtained by cutting the seed capsules of poppy flowers to collect the sticky sap
   b. Morphine and codeine are obtained from the sap extracted from the poppy
   c. Other drugs obtained from the poppy by purification are papaverine, used to treat circulatory diseases; noscapine, used as a codeine substitute; and heroin, a morphine derivative used as a pain killer
   d. Poppy seeds do not contain opium and are used as a garnish on baked products and for producing oil used in the manufacture of margarine and shortening

E. Brassicaceae (mustard family)

1. Many edible members of the mustard family are cultivated as garden vegetables: cabbage, cauliflower, brussels sprouts, broccoli, radish, kohlrabi, turnip, horseradish, watercress, and rutabaga
2. Edible weeds, such as shepherd's purse, cresses, peppergrass, and wild mustard, are also members of this family
3. The seeds of several varieties of mustards are ground to make the condiment mustard

F. Rosaceae (rose family)

1. Members of the rose family exist as trees, shrubs, and herbs
2. They produce many of the fruits included in the human diet: cherries, apples, apricots, peaches, plums, pears, strawberries, blackberries, loganberries, and raspberries
3. Roses are used as garden ornamentals, a source of oils for perfume production, and a source of vitamin C (from rose hips)

G. Fabaceae (legume family)

1. The legume family includes many species used by humans as food: peas, beans (kidney, lima, garbanzo, and broad), lentils, peanuts, licorice, carob, and soybeans
2. Other legumes, such as alfalfa and sweet clover, are used as animal feed
3. Soybeans are harvested for oil and protein
    a. Soy milk is used to make tofu and textured vegetable protein, which is used as a meat extender or substitute; soy sauce is made from the solids that remain after the soy milk is removed
    b. Soybeans are also used as cattle feed to fatten steers
4. Other useful products extracted from legumes include gum arabic and gum tragacanth, which are used to improve adhesive properties in mucilages, pastes, paints, and printing goods
5. Locoweeds, lupines, and black locusts are poisonous legumes

H. Euphorbiaceae (spurge family)

1. The spurge cassava has roots that resemble sweet potatoes and are used as food in a dried form called *farinha*
    a. Tapioca is made by forcing heated cassava through a mesh to form pellets
    b. Some cassava have cynaogenic glycosides that are toxic to humans and must be removed before consumption
2. Several members of the spurge family (most commonly, *Hevea brasiliensis*) are sources of the latex used to make natural rubber products
3. Another spurge, the castor bean plant, is both useful and toxic
    a. Castor beans are used to manufacture nylon, plastics, and soaps
    b. The seeds of ornamental castor beans contain the toxin ricin, which is extremely poisonous (one to three seeds can kill a child; four to eight can kill an adult) and produces symptoms such as mouth and throat irritation, gastroenteritis, dulled vision, extreme thirst, uremia, and sometimes death
4. The crown of thorns and the Christmas flower, or poinsettia, are ornamental spurges

I. Cactaceae (cactus family)

1. Cacti are used as houseplants; since only three varieties are poisonous, most are edible
2. Peyote cactus (*Lophophora williamsii*) is a low, spineless variety native to arid regions of Mexico and the southwestern United States
   a. This small, somewhat carrot-shaped plant grows underground with just the top exposed
   b. The top is cut off and dried to form a mescal button (or peyote button), which is consumed
   c. Peyote contains nine different alkaloid hallucinogens, the most powerful one being mescaline
   d. It is used ceremonially by native Americans for its hallucinogenic effect

J. Laminaceae (mint family)

1. Mints are generally herbs and shrubs, many of which have considerable economic importance: peppermint, spearmint, thyme, sage, oregano, marjoram, basil, catnip, rosemary, and lavender
2. Mint oils have been used as an antiseptic and as a flavoring
   a. Horehound is harvested in Europe for the production of horehound candies and cough medicines
   b. Menthol is the most abundant ingredient in peppermint oil and is widely used in toothpaste, candies, gum, liqueurs, and cigarettes
3. Peppermint is grown commercially in Oregon and Washington
4. The common houseplant coleus is a mint with colorful, attractive foliage

K. Solanaceae (nightshade family)

1. Members of the nightshade family include tomatoes, white potatoes, eggplants, peppers, tobacco, and petunias
2. Although many nightshades are poisonous, they are a source of drugs
   a. Belladonna is extracted from the leaves of *Atropa belladonna* and is the source of atropine, hyoscyamine, and scopolamine
      (1) Atropine and hyoscyamine are used to treat shock, dilate pupils for eye exams, and counteract muscle spasms; atropine is also used to treat symptomatic bradycardia
      (2) Scopolamine is a tranquilizer and is used to prevent motion sickness and as an opium antidote
   b. Capsicum is obtained from a variety of pepper and is used as a gastric stimulant
3. Tobacco (*Nicotiana tabacum*) leaves are used in cigarettes, cigars, pipe tobacco, and chewing tobacco
   a. The active ingredient of tobacco is nicotine, an addictive drug linked to throat, lung, and mouth cancers; tobacco also contains many other, more powerful carcinogens
   b. Nicotine is used as an insecticide and to kill intestinal worms in livestock
4. The white potato (*Solanum tuberosum*) is an important food crop grown in temperate areas
   a. When exposed to light, the tubers turn green on the surface—these areas are poisonous and should not be eaten
   b. The sweet potato (*Ipomoea batatas*) is not closely related to the white potato; it is a member of the morning glory family

L. Apiaceae (carrot family)
1. Useful members of the carrot family include celery, carrots, parsley, caraway, coriander, fennel, parsnip, and anise
2. Anise is used as a flavoring in cakes, pastries, candy, and liqueur (anisette)
3. Several members of the carrot family still grow wild; their roots were used by native Americans for food
4. Other members are poisonous weeds found in abandoned pastures and flood plains along streams; examples include poison hemlock and water hemlock

M. Cucurbitaceae (pumpkin family)
1. Common foods, such as pumpkins, cucumbers, squashes, cantaloupes, watermelons, and gourds, are members of this family
2. Dried luffa gourds are used as scrubbing sponges

N. Asteraceae (sunflower family)
1. Useful members of the sunflower family include sunflowers, dandelions, lettuce, endive, chicory, artichokes, chrysanthemums, marigolds, and thistle
2. Safflower and sunflower are grown commercially for oil production
3. The spice tarragon consists of the dried leaves and flowering tops of the European sagebrush, *Artemisia dracunculus*
4. Chicory, a common wildflower member of the sunflower family, is used as a coffee substitute
5. Marigolds are often planted by home gardeners to protect their crops from nematodes and insects

O. Chenopodiaceae (beet family)
1. Swiss chard, spinach, and table and sugar beets are useful members of this family
2. Spinach originated in southwestern Asia and table and sugar beets (*Beta vulgaris*) are thought to have originated in northern Europe

P. Rutaceae (citrus family)
1. Useful members of the citrus family include oranges, grapefruits, limes, lemons, and tangerines
2. Citrus fruits grow best in warm climates, such as those of Arizona, Florida, Texas, and California
3. Seedless varieties of citrus fruits are propagated vegetatively by cuttings

III. Plants That Cause Discomfort

A. General information
1. Plants produce a variety of substances to enhance their growth, prevent **herbivory**, and interfere with competitors
2. Human contact with some of these natural products often leads to discomfort in the form of hay fever or contact dermatitis

B. Hay fever
1. A sizeable percentage of the human population is allergic to pollen or spores from a variety of plants, including ragweed, corn, grasses, flowering trees, and fungi
2. Sensitive persons react to the proteins on the exterior surface of the pollen grains and spores
   a. After contact with the allergen, the human immune system increases the production and release of histamine
   b. Histamine causes the swollen membranes, runny nose, itchy eyes, and sneezing that are characteristic of hay fever; it dilates blood vessels and increases the permeability of capillaries, which results in swelling
3. Sensitivity to the allergic qualities of pollen varies significantly among individuals
4. Allergic reactions occur throughout the months that plants are growing, which can be divided into three hay fever seasons
   a. In the spring, pollen from trees (oak, hickory, elm, and box elder) is the chief irritant
   b. During the summer, pollen from grasses and late-blooming trees is problematic
   c. In autumn, ragweed and grasses are the primary sources of pollen
   d. A normally nonsensitive (nonallergic) person can become sensitized to any of these plants by constant exposure

C. Contact dermatitis
1. Contact dermatitis is characterized by irritating and painful skin rashes caused by contact with certain plants
2. Poison ivy and its relatives are the most common causes of dermatitis
   a. Poison ivy (*Toxicodendron radicans*) is a climbing vine or woody shrub that prefers moist conditions along lake shores and stream banks; it is also found at the edges of woods and around buildings
      (1) It is most readily identified by its characteristic three-leaf pattern, with the middle leaflet attached by a longer stalk
      (2) In autumn, poison ivy leaves turn bright red
      (3) The irritant is the oil in its sap, which can be carried on dust and ash
      (4) Up to 70% of the population is allergic to poison ivy to some degree
   b. Poison oak (*Toxicodendron toxicarium* and *Toxicodendron diversilobum)* are close relatives of poison ivy and are found on sandy soil and in pine forests
   c. Poison sumac (*Toxicodendron vernix*) is a shrub or tree found in bogs, along rivers, and, occasionally, along highways
3. The stinging nettle (*Urtica dioica*) is a tall perennial herb found in damp woodlands and along roadsides; its leaves and stems are covered with stinging hairs that inject a toxin when touched—this toxin causes burning and itching
4. The trumpet creeper (*Campsis radicans*) is a woody vine with red or yellow trumpet-shaped flowers and opposite, pinnately compound leaves
   a. It is found along fence posts, trees, and utility poles
   b. Contact with the plant's leaves or flowers can result in reddening of skin and formation of blisters that persist for days
5. The wild parsnip *(Pastinaca sativa)* is a common weed in fields and along roadsides; when touched, it causes inflammation, blistering, and a persistent redness that can last for weeks or months

IV. Plants as a Source of Beverages

A. General information
1. Various plant parts (seeds, leaves, and roots) have been used by humans for centuries as a source of flavoring
2. The flavoring is usually extracted from the plant part by boiling it in water, as is the case with coffee, tea, and chocolate
3. Coffee, tea, and chocolate contain caffeine or theobromine, which enhances or stimulates human metabolism

B. Coffee (*Coffea*)
1. Coffee was first grown in the Middle East approximately 1,000 years ago
2. The coffee beans, which are really the seeds, are roasted after air drying to develop flavor
3. Most of the coffee used in United States is grown in South America, with Colombia and Brazil producing the highest quality coffee
4. The principal cultivar varieties of coffee are *Coffea arabica,* which is the most common form imported by the United States and is considered to have the best flavor; and *Coffea robusta,* which has a stronger favor
5. Coffee contains 75 to 150 mg of caffeine per cup

C. Tea (*Camellia sinensis*)
1. Tea is the second most-consumed beverage in the world (water is number one)
2. Tea plants are native to China and northern India; most of the tea exported to the United States is from India
3. Of the many types of teas, Earl Grey, black tea, and green tea are the most common
   a. Earl Grey tea is flavored with oil of bergamot, a member of the citrus family
   b. Black tea is the most common bagged tea
      (1) It is harvested as newly opened leaves that are air dried to allow flavor to develop, then heated to deactivate enzymes and further develop the flavor
      (2) As the leaves are air dried and heated, they turn black
   c. Green tea leaves are steamed before crushing and drying
      (1) Steaming prevents the leaves from darkening
      (2) The green leaves produce a milder flavor
4. Approximately half of the tea consumed in the United States is in the form of iced tea
5. Tea bags were accidently invented in 1908 by Thomas Sullivan, a tea importer-wholesaler from New York who sent samples of tea packaged in small silk bags to prospective customers; one of the customers brewed the tea without removing it from the bag
6. Tea contains slightly less than 50 mg of caffeine per cup

D. Chocolate
1. Chocolate is derived from the cacao plant (*Theobroma cacao*) and was first used 4,000 years ago by native peoples of the Amazon River basin
2. Much of today's chocolate comes from Brazil and the Ivory Coast, with smaller amounts from Venezuela, Ecuador, Nigeria, and Ghana

3. The cacao plant has a large pod (10 to 12 cm long) that contains 40 seeds, or beans
   a. The seeds are fermented in large vats to develop color, aroma, and flavor
   b. They are then ground under heat and pressure to make a fluid, called chocolate liquor, that is formed into the hardened bars known as baking chocolate
   c. About 40% (by weight) of the cacao bean is fat (cocoa butter)
      (1) The cocoa butter is removed from the liquor by pressure, leaving a chocolate "press cake" that is ground into cocoa powder
      (2) Milk chocolate is made by mixing milk, cocoa butter, and cocoa powder
      (3) Other edible chocolate varieties are produced by mixing various amounts of cocoa butter and cocoa powder
4. Chocolate contains 50 mg of theobromine per cup of hot chocolate

E. Alcoholic beverages
1. The ethyl alcohol in alcoholic beverages is produced by the fermentation of sugar from plant products
2. The fermentation process is triggered by yeasts, generally those from the genus *Saccharomyces*
3. Wines are generally made by fermentation of grape juice
   a. A wine's flavor is determined by the type of grape, the conditions of cultivation, and the fermentation processes
   b. Grapes that produce the best wines are native to the Mediterranean region, but many cultivars are grown in other regions of the world as well
   c. Wines are named for the growing region or the type of grape used in their production; wines named for grape type are called varietal wines
4. Beer is produced by fermentation of starchy materials from grains
   a. Many different grains, including rice, corn, barley, hops, and wheat, can be used to supply the starch
   b. Malt, which is formed from germinating barley seeds, provides the enzymes that digest the starch into sugars for use by yeast cells
   c. A beer's flavor is significantly affected by the addition of inflorescences from the plant hops (*Humulus lupulus*)

V. Plants as a Source of Fiber

A. General information
1. Humans have been using plant fibers for textiles and ropes since before recorded history
2. Plant fibers can be blended with synthetic fibers to make a variety of materials

B. Cotton
1. Cotton is the most important plant fiber because of its tensile strength, ease with which it accepts dyes, and thermal characteristics
2. Cotton fibers are derived from the surface hairs on the seeds of the cotton plant (*Gossypium*); these surface hairs develop while the seeds are in a pod or capsule
3. Cotton fibers are more than 90% cellulose and are used to make fabrics and clothing

C. Flax

1. Flax fibers are obtained from the stem of the flax plant (*Linum usitatissimum*)
2. Flax was once the world's leading soft fiber, but has been replaced by cotton
3. Commercial production of flax is concentrated in northern Europe
4. Flax fibers are used to make linen, and flax seeds are used to produce linseed oil

D. Jute

1. Jute fibers are obtained from *Corchorus capsularis* and *Corchorus olitorius*, plants native to southeast Asia
2. These fibers are used in the production of burlap and some twines

E. Manila hemp (abaca)

1. Hemp fibers are obtained from an inedible banana (*Musa textilis*), which is native to the Philippines
2. Hemp is used to manufacture rope and twine

VI. Plants as a Source of Drugs

A. General information

1. The history of medicine and botany are closely intertwined, since many early physicians were part-time botanists and herbalists
2. Primitive cultures used plants extensively in a trial-and-error process to treat diseases
3. Plant extracts have served as starting points for the discovery of numerous pharmaceutical items
4. Many drugs that were originally extracted from plants are now produced synthetically

B. Opium

1. Humans have used opium, a drug extracted from the resins in the sap of the poppy (*Papaver somniferum*), for more than 6,000 years
2. Most poppies are grown in India, Pakistan, Afghanistan, Iran, Turkey, Burma, Thailand, and Laos
3. Opium derivatives include morphine, heroin, and codeine
    a. Morphine and heroin are legally used as pain relievers, but are psychologically and physically addictive
    b. Codeine suppresses the cough center and is an ingredient in some cough medicines

C. Cortisone

1. Yam tubers (*Dioscorea*) from the tropics serve as a source (directly or as the raw material dioscin) of medically significant steroids, including cortisone, testosterone, estrogen, and progesterone
2. Cortisone is commonly used in skin creams to reduce itching and as an anti-inflammatory drug

D. Other plant-derived drugs

1. Colchicine is extracted from the corms of autumn crocus (lily family) and used in the treatment of gout and rheumatism and in the preparation of chromosomes in genetic studies of plants and animals

2. Podophyllin is extracted from mayapple (*Podophyllum peltatum*) and used to treat brain tumors

3. Digitalis is extracted from foxglove (*Digitalis purpurea*) and is useful in strengthening cardiac muscle

4. Quinine is extracted from the bark of the *Cinchona* tree (native to South America) and is used to treat malaria and as a flavoring in tonic water

5. Reserpine is extracted from the root of the snakeroot plant (*Rauwolfia serpentina*) and is used to treat schizophrenia (as a tranquilizer) and high blood pressure

6. Taxol is extracted from the bark of the Pacific yew tree and recently has been successfully used in the treatment of breast cancer

7. Ephedrine is extracted from the gymnosperm shrub *Ephedra* and used to relieve allergic symptoms

---

Study Activities

1. List two monocot and eight dicot plant families and discuss their importance to humans.
2. Name the four sources of plant fibers used by humans.
3. List six drugs and the plants from which they are derived.
4. Using the information in this chapter, prepare a list of the plants that you use directly and indirectly in one day.

16

Plant Ecology

Objectives

After studying this chapter, the reader should be able to:
- Explain the terms ecology, population, community, ecosystem, and biome.
- Give examples of mutualism in plants.
- Describe allelopathy.
- Describe plant defense mechanisms against herbivores.
- Describe the flow of energy through an ecosystem.
- Explain ecologic succession.
- Name and describe the terrestrial biomes.

I. Populations, Communities, and Ecosystems

A. General information
1. *Ecology* is the study of organisms and their interactions with one another and the environment
2. Ecologists study different realms, or levels of complexity
   a. A *population* is a group of individuals of the same species that live in a particular geographic area
   b. A *community* is a collection of several interconnected populations in a specific area
   c. An *ecosystem* is the relationship of the biologic community to its physical environment; it is composed of *abiotic* (nonliving elements, such as soil, water, air, temperature, and nutrients) and *biotic* (living organisms, such as plants and animals) resources
   d. A *biome* is a large, expansive ecosystem that covers a huge geographic area or climatic zone

B. Interactions
1. Organisms within a particular environment develop mutual or competitive relationships
2. *Mutualism* describes cooperative interactions among organisms in which the growth and survival of both interacting species are enhanced
   a. One example of mutualism is the mycorrhizal relationship between fungi and vascular plants, which result in better growth and nutrition for each organism

   b. Another example is the mutualistic association between acacia trees and
      ants
      (1) The ants protect the tree from herbivores and remove adjacent vegeta-
          tion that could interfere with the tree's growth
      (2) In return, the ants feed on the tree's *nectaries* (glands that secrete nec-
          tar) and often live in its hollow thorns on branches
3. **Competition** is an interaction in which organisms are adversely affected
   a. Organisms compete for resources that are in short supply and used by other
      organisms simultaneously
   b. *Intraspecific competition* is an intense form of competition between members
      of the same species; *interspecific competition* is a less intense form be-
      tween different species
   c. Plants compete for nutrients, water, and sunlight
   d. Competition among plants affects growth rates, leaf arrangement, and over-
      all plant shape
   e. Plants that do not compete successfully may become stunted or weakened
   f. In experimental conditions, two species that are growing together and using
      the same resources cannot coexist indefinitely—one species is eventually
      eliminated; this principle is called **competitive exclusion**
   g. In natural conditions, species that grow together may use resources in
      slightly different ways, allowing them to coexist
   h. Competitive situations are variable and complicated and often difficult for re-
      searchers to quantify or prove
   i. **Allelopathy** is a type of competition in which plants produce chemicals that
      inhibit the growth or germination of competing species; ferns, black wal-
      nuts, sunflowers, and creosote bushes are allelopathic plants
4. *Herbivory* is an interaction in which animals consume plants or plant parts for
   food
   a. Plants discourage herbivores with chemical defenses (alkaloids, tannins, ter-
      penes, and phenolics) or mechanical defenses (thorns, spines, thick bark,
      and hairs)
   b. Chemical defenses are complex and varied
      (1) Oak trees (*Quercus*) defoliated by gypsy moths (*Lymantria dispar*) pro-
          duce new leaves with a higher tannin and lower water content, which
          allows the leaves to inhibit the growth and survival of future genera-
          tions of moth larvae
      (2) Paper birch (*Betula papyrifera*) increases its production of resins and
          phenolic compounds, which make the plant less palatable to such her-
          bivores as the snowshoe hare
      (3) Wild-growing members of the squash family produce terpenes, which
          have a bitter taste that prevents herbivores from eating the leaves and
          fruits

C. Nutrient cycling
   1. Nutrients are readily recycled among organisms within an ecosystem
   2. Plants play a prominent role in the retention of nutrients; for example, ecosys-
      tems encompassing deciduous forests are very efficient in conserving mineral
      elements

3. Few nutrients escape from intact or undisturbed ecosystems; however, large amounts of nutrients may be lost from disturbed ecosystems (those harmed by fire, clear-cut logging, or construction), possibly leading to soil erosion

4. The most important **biogeochemical cycles** are those for water, nitrogen, phosphorus, and carbon

   a. The **hydrologic** (global water) **cycle** describes the movement of water from atmosphere to earth

     (1) The movement of water in the hydrologic cycle is driven by solar energy

     (2) Water moves from the atmosphere to the earth's surface as precipitation; it returns to the atmosphere by evaporation from bodies of water and transpiration from plants

       (a) In some terrestrial ecosystems, more than 90% of the water returns to the atmosphere during plant transpiration

       (b) The remaining 10% evaporates directly from substrate surfaces

     (3) Through transpiration, plants have a substantial role in returning water to the atmosphere and thus can affect climatic conditions

     (4) The ongoing and massive destruction of forest ecosystems has the potential to change the climate on a global scale by altering the flow of water through the hydrologic cycle

   b. The *nitrogen cycle* describes the conversion of nitrogen in the atmosphere to biologically usable forms (see *Nitrogen Cycle*, page 124)

     (1) All organisms contain large amounts of nitrogen in the form of proteins and nucleic acids

     (2) Although nitrogen gas makes up 79% of the earth's atmosphere, most organisms cannot use nitrogen in its gaseous form

     (3) Nitrogen fixation, a process that converts atmospheric nitrogen to biologically usable forms, can result from the actions of lightning, volcanoes, or nitrogen-fixing bacteria that live in the soil or root nodules of certain plants

     (4) During nitrogen fixation, nitrogen gas (N_2) is converted to ammonia (NH_3), which dissolves in water to form a biologically usable ammonium ion (NH_4^+)

     (5) Plants can absorb the products of nitrogen fixation, change them into nitrate (NO_3^-), and incorporate them into proteins, nucleic acids, and other nitrogen-containing compounds

     (6) Animals that eat plant matter break down the nitrogen-containing compounds and use them to form new animal proteins

     (7) Dead plants and animals provide decomposer bacteria with sources of nitrogen, which is used in the production of ammonia (a process called *ammonification*)

     (8) Other bacteria change the ammonia to nitrate (NO_3^-) in a process called **nitrification**

     (9) Still other bacteria act on the remaining nitrate, changing it back into nitrogen gas in **denitrification**, a process that completes the nitrogen cycle

   c. The *phosphorus cycle* consists of two interconnected (local and global) circuits

     (1) In the local circuit, phosphorus moves from rocks and soil to organisms and back to the soil

Nitrogen Cycle

Nitrogen fixation by bacteria in soil, water, or root nodules or from lightning and volcanic activity converts atmospheric nitrogen to biologically useful forms. Plants assimilate the nitrogen and incorporate it into proteins and other molecules, which are converted to animal proteins when the plants are consumed. Ammonia from dead, decaying plants and animals is usually converted to nitrate or absorbed directly by plants. Denitrifying bacteria convert biologically useful forms of nitrogen into nitrogen gas.

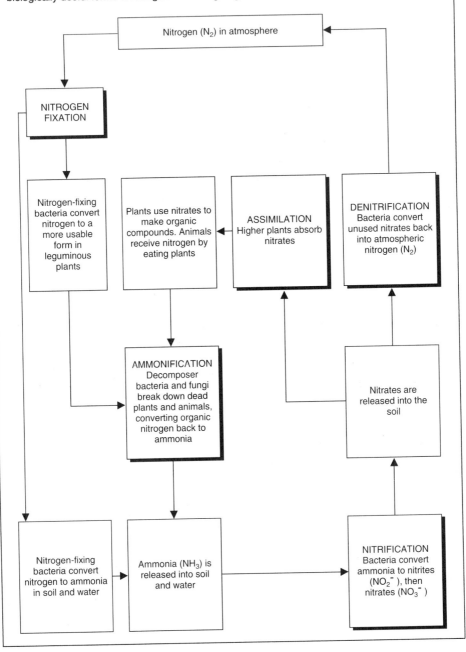

        (a) Plants absorb phosphate (PO_4^{-3}) directly from the soil or water; animals obtain phosphate by eating the plants or other animals

        (b) When the plants and animals die and decompose, the organic phosphorus in their tissues is converted by bacteria into phosphate, which returns to the soil, and the cycle begins again

      (2) In the global circuit, phosphorus leaves local ecosystems by way of streams and rivers

        (a) Phosphorus transported in this fashion is incorporated into ocean sediments and may remain unavailable to terrestrial ecosystems for long periods of time

        (b) Phosphorus trapped in ocean sediments becomes available when the sea floor rises and exposes the phosphorus-containing materials

    d. The *carbon cycle* is connected to the flow of energy within ecosystems

      (1) Carbon, in the form of carbon dioxide, is trapped in sugars during photosynthesis

      (2) During cellular respiration, the carbon within the sugars is returned to the atmosphere as carbon dioxide

      (3) Some carbon accumulates in wood and other tissues for many years; it eventually returns to the atmosphere when released by fire or through consumption and respiration by fungi and bacteria

      (4) Carbon can leave the cycle for long periods of time when buried as organic litter, which decomposes only partially and is transformed by pressure and heat into coal and petroleum

D. Energy flow and trophic levels

    1. Energy within most ecosystems ultimately comes from the sun; consequently, plants play a vital role in converting solar energy to chemical energy through photosynthesis

    2. As plants are consumed by animals, the energy is transferred along the **food chain**, which is more correctly described as a **food web**

    3. The various feeding levels along the food chain are called **trophic levels** (see *Trophic Structure*, page 126)

    4. Each trophic level serves as a food source for the next higher trophic level

      a. **Producers** (plants) are the foundation of many food chains

      b. Primary consumers (herbivores) eat producers; a grasshopper is a primary consumer

      c. Secondary consumers (**carnivores**) eat primary consumers; a bird that eats the grasshopper is a secondary consumer

      d. Tertiary consumers eat other carnivores; a weasel that eats an insect-eating bird is a tertiary consumer

      e. **Detritivores** are a special class of consumers that obtain energy and materials from the **detritus** which accumulates from all trophic levels

        (1) Bacteria and fungi are the most common detritivores

        (2) This group also includes worms, nematodes, some insects, and animals (such as vultures) that feed on carrion

    5. In most ecosystems, energy flows through autotrophs to consumers to decomposers

    6. Much of the energy transferred from one trophic level to the next is lost as heat or cannot be harvested by the next trophic level

Trophic Structure

Feeding arrangements of ecosystems, commonly called food chains, are typically more complex and thus more correctly referred to as food webs. At each step in the transfer of energy from one trophic, or feeding, level to the next, much energy is lost.

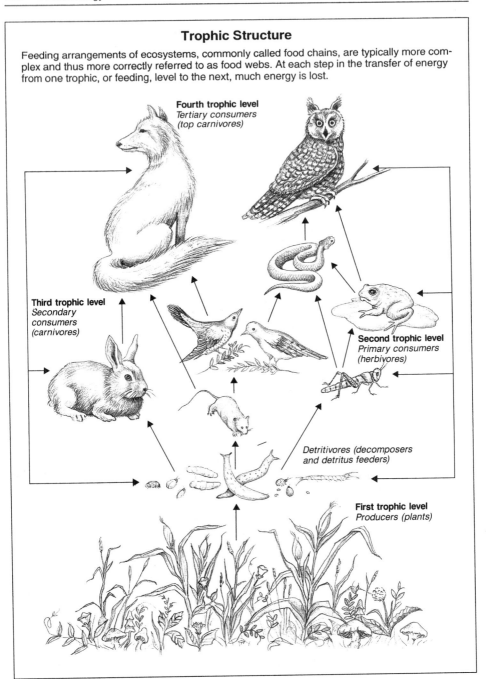

Fourth trophic level
Tertiary consumers
(top carnivores)

Third trophic level
Secondary
consumers
(carnivores)

Second trophic level
Primary consumers
(herbivores)

Detritivores (decomposers
and detritus feeders)

First trophic level
Producers (plants)

a. Only about 2% to 10% of the usable energy in one trophic level is transferred to the succeeding level

b. Food chains are generally limited to three or four levels because of the inefficiency of energy transfer

E. Succession

1. Ecologic *succession* is the orderly development of a biologic community in a particular area
2. It is characterized by progressive changes in the community over time
   a. A *pioneer community* is the first plant community to enter an area
   b. *Successional communities* are transitions between pioneer and climax communities
   c. A *climax community* is a self-perpetuating, more or less steady-state community; geographic areas are characterized by climax communities, which are determined by physical and climatic conditions
3. Succession occurs at a variable rate in all temporarily open areas
   a. *Primary succession* occurs on new land (created by volcano flows, newly uplifted land masses, or glacial tills) or previously barren land
      (1) The early stages of primary succession involve soil formation and enrichment by the addition of organic materials to the substrate, as plants or plant parts die and decompose
      (2) Because the land is barren, plants arrive (often as seeds carried by wind, water, or animals) from other areas to colonize the new habitat
   b. *Secondary succession* is a "healing" process that occurs after a community has been disturbed (for example, by farming, logging, or fire)
      (1) Secondary succession occurs more quickly than primary succession because seeds and organisms are readily available
      (2) Unlike primary succession, soil is already formed and enriched, which aids in the establishment of the new community
4. Succession increases the amount of organic matter in the soil, promotes diversity in the number and variety of species within the community, and decreases the amount of nutrients lost during nutrient cycling (because more nutrients are held within the vegetation)

II. Terrestrial Biomes

A. General information

1. Terrestrial biomes (as opposed to aquatic biomes) are determined largely by temperature and moisture
2. Terrestrial biomes are classified according to the predominant plant types and include tundra, taiga, temperate deciduous forest, savanna, desert, tropical rain forest, chaparral, and temperate grassland

B. Tundra

1. The tundra marks the northernmost limit of plant growth and occupies approximately 20% of the earth's surface
2. The arctic tundra ranges from the North Pole to the northernmost edge of the taiga; the alpine tundra is located above the tree line on high mountains
3. The arctic tundra is characterized by permafrost, a continuously frozen ground layer that restricts root growth and water movement within the soil
4. Plant forms are limited to low, shrubby or matlike vegetation due to the cold temperatures and wind patterns; these plants exhibit slow rates of photosynthesis all year round

5. The predominant forms of tundra vegetation are shrubs, sedges, grasses, mosses, and lichens; miniature willow (*Salix*) and birch (*Betula*) trees also grow, although they reach heights of only 1" to 20" (2.5 to 50 cm)
6. The brief growing season of two to three months (although frost can occur on any day) is marked by nearly 24 hours of daylight, during which perennial plants often form brilliant flowering mats in meadow areas
7. Because of the harsh environment and thin topsoil (only 2" to 3" [5 to 7.5 cm]), the tundra tends to be a fragile biome; the effects of any disturbance are likely to remain for long periods
8. Animals of the tundra include lemmings, arctic foxes, snowy owls, caribou, polar bears, shrews, ptarmigans, loons, plovers, jaegers, and arctic terns

C. Taiga (northern coniferous or boreal forest)

1. The taiga extends across North America, Europe, and Asia, and northward to the arctic tundra
2. It is characterized by harsh winters (temperatures can drop to -58° F [-50° C]) and brief but warm (68° to 86° F [20° to 30° C]) summers
3. The considerable precipitation occurs mostly in the summer; snow covers the area in winter
4. Taiga soils tend to be thin and acidic and often poor in mineral nutrients; the soils in the northern portions of the taiga have a layter of permafrost beneath the topsoil
5. Conifer species of the northern boreal forest typically include spruce, pine, fir, and hemlock
6. Animals of the taiga include shrews, jumping mice, moose, ermines, wolverines, caribou, and various birds

D. Temperate deciduous forest

1. The temperate deciduous forest is located in middle latitude regions with suffi- cient moisture to support large trees; the eastern United States, most of middle Europe, and parts of eastern Asia fit this description
2. Temperate forests are characterized by broad-leaved deciduous trees
3. Temperatures are cold in winter (below 39° F [4° C]) and hot (above 86° F [20° C]) in summer
4. The relatively abundant precipitation is fairly evenly distributed throughout the year, with slightly more precipitation occurring in the summer
5. During the summer, the trees of the deciduous forest form a nearly solid canopy that shades the forest floor from direct sunlight
   a. The shading effect of the forest canopy results in a progression of flowering throughout the growing season
   b. In the spring, before the canopy is fully formed, spring flowers on the forest floor bloom
   c. In midsummer through fall, plants that can tolerate more shade bloom
6. Animals of the temperate deciduous forest include foxes, raccoons, opossums, squirrels, snakes, mice, salamanders, and a variety of birds

E. Savanna

1. Located in central South America, central and south Africa, and parts of Austra- lia, the savannas are grasslands with scattered, individual trees

2. Savannas are generally tropical and subtropical regions with three seasons—cool and dry, hot and dry, and warm and wet
3. Trees of the savanna are generally deciduous and lose their leaves during the dry seasons
4. Fires are a frequent occurrence in this biome
   a. Many of the grass species rapidly recover after a fire
   b. Because the trees are less fire-tolerant, forests do not exist
5. Animals of the savanna include giraffes, zebras, antelopes, buffalo, kangaroos, mice, moles, gophers, snakes, ground squirrels, and many arthropods

F. Desert
   1. Deserts can be hot or cold, depending on the prevailing temperature; they occur wherever precipitation is consistently low or the soil is too porous to retain water
   2. Most deserts receive less than 5" (12.5 cm) of precipitation per year
   3. They are characterized by considerable shifts in daily temperature; air temperatures can range from more than 97° F (36° C) during the day to below 59° F (15° C) at night
   4. Hot deserts are found in North America, South America, North Africa, the Middle East, and Australia; cold deserts are located in North America, Argentina, and central Asia
   5. The life cycles of desert plants are keyed to rainfall, not temperature, as evidenced by the large blooms that occur immediately after the infrequent rainstorms
   6. Many desert plants have adapted to the desert's high light intensities through evolution of Crassulacean acid metabolism (CAM) photosynthesis
      a. CAM plants open their stomata at night to allow gas exchange and close them during the day to prevent excessive water loss
      b. CAM metabolism permits more photosynthesis to take place than would otherwise be possible in a desert environment
   7. Desert animals are active at night when lower temperatures prevail and include mice, kangaroo rats, snakes, chuckwallas, lizards, and various birds

G. Tropical rain forest
   1. Because tropical rain forests are located near the equator, the day length (12 hours of light per day) varies little throughout the year
   2. Temperatures also remain constant year round, generally ranging from 77° to 95° F (25° to 35° C)
   3. Rainfall is significant—between 79" to 156" (200 to 400 cm) per year; most rain forests have no dry season
   4. The rainfall and temperature conditions support a diversity of flora and fauna; rain forests have more species than all other biomes combined
   5. The dense rain forest canopy prevents most light from reaching the forest floor, resulting in keen competition for light by plants
   6. Many rain forest trees are covered by *epiphytes*, such as orchids and bromeliads
   7. Despite the abundance of vegetation, little litter or humus accumulates, and the soils are generally poor in nutrient content
   8. Decomposition occurs rapidly, and the nutrients from decomposing materials are quickly recycled into the plant community or leached away by the heavy rains

9. Animals of the tropical rain forest include tree frogs, monkeys, sloths, snakes, lizards, numerous insects, and a large variety of birds

H. Chaparral
1. Chaparrals are regions of dense, spiny shrubs with tough evergreen leaves
2. They are located in middle latitude areas along the coasts of California, Chile, southwestern Africa, and southwestern Australia
3. Fire plays an important role in maintaining the characteristic vegetation of the chaparral
    a. Chaparral shrubs have root systems adapted to fire or produce seeds that germinate after exposure to the heat of a fire
    b. Competitor plants, such as grasses and trees, cannot recover as readily as the shrubs can from a fire
4. Animals of the chaparral include deer, fruit-eating birds, rodents, snakes and lizards

I. Temperate grassland
1. Temperate grasslands are located in the temperate areas of South Africa, Argentina, North America, and extend from the western portion of the former Soviet Union into the Asian continent
2. Grasslands are similar to savannas but occur in colder regions
3. They are perpetuated by periodic fires and drought that prevent the permanent establishment of trees and shrubs
4. They tend to intergrade (merge gradually) with the forests, woodlands, and deserts that exist at their margins
5. Grassland animals include rabbits, gophers, mice, pronghorns, buffalo, and numerous birds

Study Activities

1. Describe the relationships between population, community, ecosystem, and biome.
2. Name three chemical and two mechanical defense mechanisms used by plants to prevent herbivory.
3. Explain the important nutrient cycles in ecosystems.
4. Discuss the components of the food web.
5. List and describe the terrestrial biomes.

17

Plant Genetics

Objectives

After studying this chapter, the reader should be able to:
- Define gene, allele, dominance, phenotype, genotype, and linkage.
- Understand the importance of Mendel's experiments.
- Understand the inheritance patterns of monohybrid and dihybrid crosses.
- Explain Mendel's laws of inheritance and describe exceptions to them.
- Explain the basic methods of plant breeding.

I. Introduction to Genetics

A. General information
  1. *Genetics*—the study of inheritance—began with Gregor Mendel, an Austrian monk who studied heredity in garden peas
  2. Published in 1866, his work was largely overlooked until the early 1900s
  3. Classic or Mendelian genetics is the study of the inheritance of the physical characteristics of an organism
  4. Molecular genetics is the study of the inheritance of specific proteins and the metabolic activity of these proteins, which, in turn, determine the physical characteristics of an organism

B. Mendelian genetics
  1. Before Mendel, knowledge of how characteristics are passed from parent to offspring was limited
  2. Mendel's genetic studies were conducted in garden peas, an ideal plant for such work because they are normally self-pollinated, and thus all the seeds from the same plant have the same inherited characteristics
  3. In his first experiments, Mendel crossed (mated) tall plants with dwarf plants
     a. He opened the flowers, removed the anthers to prevent self-pollination, manually transferred the appropriate pollen from plant to plant, then planted the seeds
     b. The offspring (called the first *filial*, or F1, generation) were all tall plants
     c. By allowing the plants of the F1 generation to self-pollinate, Mendel produced a second (F2) generation of three tall plants and one short plant
  4. The results of his first experiments led Mendel to conclude that two hereditary units (which he called *factors*) control each physical characteristic or trait

      a. He arrived at this conclusion by observing that dwarfism disappeared in the F1 generation but reappeared in the F2 generation

      b. If the plants had only one hereditary factor, dwarfism could not have reappeared in the F2 generation

5. Based on his experiments, Mendel devised a set of rules governing inheritance; these rules are known as Mendel's laws

    a. The *law of unit factors* states that factors always occur in pairs and control the inheritance of various characteristics or traits of an organism

      (1) The paired factors are known as ***alleles*** and are considered to be alternative molecular forms of a ***gene***

      (2) Genes are the units of instructions that produce or influence a specific trait or characteristic

    b. The *law of dominance* states that in any given pair of alleles, one is dominant and the other is recessive; the ***dominant allele*** may suppress or mask the expression of the ***recessive allele***, as evidenced by the dominance of the tall allele over the dwarf allele in the pea plants

    c. The *law of segregation* states that during the formation of gametes (egg and sperm) in sexual reproduction, the homologous chromosomes on which alleles are located are separated during meiosis and do not reunite until fertilization occurs; this separation of the alleles is called segregation

    d. The *law of independent assortment* states that the alleles controlling two or more pairs of characteristics segregate or separate independently (randomly) and that the gametes formed during meiosis combine randomly during fertilization

6. The two alleles are carried on homologous chromosomes

    a. In a ***homozygous*** organism, the chromosomes of the diploid nucleus carry the same allele for a trait (for example, both carry the allele for "tall")

    b. In a ***heterozygous*** organism, the diploid nucleus contains two different alleles (for example, one carries the allele for "tall" and the other for "dwarf")

7. The ***phenotype*** of an organism refers to its physical appearance

8. The ***genotype*** refers to the genetic information controlling physical appearance

    a. The dominant allele of the genotype is designated by a capital letter, and the recessive allele is designated by a lowercase letter

    b. In Mendel's work, the tall parent genotype would be TT and the dwarf parent genotype would be tt

C. Monohybrid Cross

1. A ***monohybrid cross*** is the mating of two parents that differ in only a single trait

2. The concepts of a monohybrid cross can be understood by examining Mendel's original cross between a homozygous tall plant (genotype TT) and a homozygous dwarf plant (genotype tt)

    a. In this cross, or mating, the resulting offspring receive one allele from the tall parent (T) and the other allele from the dwarf parent (t)

    b. All members of the F1 generation have a heterozygous genotype (Tt), and, because the tall allele is dominant, they have a "tall" phenotype

    c. When members of the F1 generation are crossed (F1 × F1), two types of gametes are produced by these heterozygous (Tt) parents—half are T and the other half are t

d. When the gametes of the F1 parents unite to form the F2 generation, three outcomes are possible: homozygous dominant (TT), heterozygous (Tt), and homozygous recessive (tt)

    (1) The genotypes in the F2 generation are produced in a ratio of 1 TT: 2 Tt: 1 tt

    (2) The phenotypes in the F2 generation would be tall in three-fourths of the individuals and dwarf in one-fourth

D. Dihybrid cross

1. A **dihybrid cross** is the mating of two parents that differ in two traits
2. Homologous chromosomes contain numerous alleles each defining different characteristics or traits
3. Different sets of genes located on different chromosomes are separated during meiosis and combined again during fertilization, as described by Mendel's law of independent assortment
4. The concept of a dihybrid cross can be understood by examining a cross between pea plants with different alleles governing height (tall, dwarf) and flower color (yellow, green)

    a. A cross of a homozygous tall yellow plant with a homozygous dwarf green plant is represented by: TTYY × ttyy

    b. All the gametes from the dominant parent will be TY; all of those from the recessive parent will be ty

    c. All members of the F1 generation will be TtYy

    d. When members of the F1 generation are crossed (F1 × F1), four types of gametes are produced because of independent assortment—TY, Ty, tY, ty

    e. When the gametes from each F1 parent unite to form the F2 generation, 16 different combinations are possible

        (1) The genotypes of the F2 generation are produced in a ratio of 1 TTYY: 2 TTYy: 2 TtYY: 4 TtYy: 1 TTyy: 2 Ttyy: 1 ttYY: 2 ttYy: 1 ttyy

        (2) The phenotypes of the F2 generation are produced in a ratio of 9 tall yellow: 3 tall green: 3 dwarf yellow: 1 dwarf green

E. Linkage

1. **Linkage** is the tendency of genes on the same chromosome to be inherited together; the genes inherited as a group are called *linkage groups*
2. Each chromosome contains many genes; the closer the genes are to one another on a chromosome, the more likely they will be inherited together
3. As a result of linkage, two or more traits may stay together from one generation to the next
4. Linkage groups can become separated in meiosis during prophase I when the homologous chromosomes align with one another and exchange genetic material (called *crossing over*)
5. The closer that two genes are on the chromosome, the less likely that they will be separated during crossing over; information gained from cross-linked genes is used by geneticists to map the position of genes on a chromosome

F. Variations on Mendel's concepts

1. In **incomplete dominance**, the offspring have a phenotype that reflects both the dominant and recessive traits; for example, the mating of a white-flowered snapdragon with a red-flowered snapdragon produces pink-flowered offspring

2. The external environment can affect gene expression and therefore change the phenotype; for example, seedlings grown in darkness may possess the genes for chlorophyll synthesis but will not do so until exposed to sunlight
3. In continuous variation, phenotypic differences within a population may exhibit a broad range of expression; for example, plants display a range of heights rather than discrete or distinct height classes

G. Mutations

1. *Mutations* are a change in a portion of a gene's deoxyribonucleic acid (DNA) that results in a change in the protein product of that gene; the altered (mutated) allele is inherited just like any other allele
2. The rate at which mutations occur is a function of the organism and the gene itself
3. Mutations can occur because of exposure to some physical or chemical agent, known as a *mutagen,* or they may appear without any attributable cause (called *spontaneous mutations*)
4. Mutations introduce new and novel genes into a population and are the ultimate source of these new genes
5. Most mutations are considered harmful and generally diminish the possibility of an individual's survival under current environmental conditions

H. Cytoplasmic inheritance

1. Plant characteristics are sometimes inherited via genes located in the cytoplasmic structures, rather than the nucleus
2. Cytoplasmic genes are found in plastids (such as chloroplasts) and mitochondria, which contain unique, circular chromosomes
   a. During sexual reproduction in many plants, the fusion of male and female gametes involves the nuclei only; because the cytoplasm of the male gamete does not enter the egg cell, it is not passed on to the zygote
   b. Thus all the plastids and mitochondria of the offspring are received from the female via the egg cytoplasm
3. Typically cytoplasmic genes are of no consequence to plant breeders, but are of interest to plant physiologists studying cell metabolism

II. Plant Breeding

A. General information

1. Most plants used by humans have been developed from ancestral plants through controlled plant breeding
2. Plant breeding can be used to reduce world hunger by improving crop yields, increase plant resistance to disease, and exploit plants as sources of new foods, medicines, and fibers
3. The basic methods used in plant breeding are hybridization, polyploidy, mutation, and tissue culture

B. Hybridization

1. *Hybridization* is the crossing (mating) of different varieties of plants
   a. Hybrid plants of the F1 generation often display positive effects of the cross
   b. These plants may be bigger, stronger, have higher yields, or be more resistant to disease

2. *Hybrid vigor* describes the increased vigor of the F1 offspring; the effect usually disappears in the next (F2) generation
3. During hybridization, plant reproduction is controlled to reinforce the desirable traits of one parent or both parents
   a. Plants may be repeatedly self-pollinated or cross-pollinated to obtain the desired characteristics
   b. As a result of careful selection, desirable dominant and recessive traits may be retained
4. Most of the corn grown in North America comes from F1 hybrid seed, and many wheat varieties are hybridized to help them resist fungal infection

C. Polyploidy
1. *Polyploidy* is a condition in which the nucleus contains more than the normal two sets of chromosomes; polyploids have three, four, six, or more sets
2. It can arise through hybridization or during meiosis or mitosis, if the diploid set of homologous chromosomes fails to separate in a gamete
3. Polyploid plants are often more vigorous than their diploid counterparts
4. Polyploidy can be artificially induced by treating seeds with colchicine, which interferes with mitosis and prevents the chromosomes from separating

D. Mutation
1. To improve plant stock, plant breeders may induce a mutation or search for a naturally occurring mutation
2. Although most mutations are harmful, some useful modifications have been obtained through induced mutations, including blight-resistant varieties of oats and improved varieties of grapes

E. Tissue culture
1. Tissue culture is the growing of living tissue in an artificial growth medium
   a. Tissue cultures can be started from nearly any plant part, but the cells near meristems usually produce the best results
   b. Within the proper media, the plant cells can form shoots and roots
2. Plant breeders typically subject the cells of a tissue culture to various stresses, such as heat, cold, herbicides, disease organisms, or radiation
   a. The cells that survive can be induced to form plants that may be resistant to some of the stresses
   b. The cells also may be subjected to genetic engineering techniques in which the genes are manipulated to produce novel gene combinations
3. Tissue cultures have been used to grow the tissues of carrots, tobacco, citrus, and other plants for use in physiologic experiments and plant breeding

Study Activities

1. Outline Mendel's experiments and discuss the laws of inheritance derived from them.
2. Describe gene linkage and explain how crossing over affects the inheritance of linked genes.
3. Describe the methods used by plant breeders to improve plant species.

Appendix

Selected References

Index

Appendix: Glossary

Abiotic—nonliving elements in an environment

Abscisic acid—growth-inhibiting hormone that promotes dormancy in plants

Abscission—separation and dropping off of plant parts (leaves, fruits, flowers) after formation of an abscission zone; results in loss of plant parts at the end of the growing season or when parts are no longer functional

Abscission initials—cluster of cells at the base of a leaf, fruit, flower, or other plant part that remains dormant until abscission

Abscission zone—area at the base of a leaf, fruit, flower or other plant body that contains abscission initials

Accessory pigment—pigment that captures light energy and transfers it to chlorophyll a

Active transport—energy-requiring transport of a solute across a membrane in a direction of increasing concentration

Adenosine triphosphate (ATP)—molecule consisting of a nitrogen base, sugar, and three phosphate groups; the high-energy chemical bonds holding the phosphate groups absorb energy when formed and release energy when broken

Aerobic respiration—oxygen-requiring pathways of fuel breakdown that release energy and take place within the mitochondria of eukaryotic cells

Aggregate fleshy fruit—fruit derived from a single flower with many pistils

Allele—alternate form of a gene

Allelopathy—inhibition of one plant by another through the production and secretion of one or more chemicals

Alternation of generations—method of reproduction in which a haploid gametophyte generation alternates with a diploid sporophyte generation; the gametophyte generation produces spores by meiosis, the sporophyte generation produces gametes by mitosis

Amino acid—organic molecules serving as the monomers for proteins or polypeptides

Ammonification—form of bacterial metabolism in which nitrogen compounds are converted to ammonia

Amyloplast—colorless plastid found in plant cells that forms starch grains

Anaerobic respiration—pathways of fuel breakdown that release energy but do not require oxygen

Angiosperm—flowering plant whose seeds develop within ovaries that mature into fruits

Annual—plant that completes its life cycle in one growing season

Annulus—in ferns, a row of specialized cells of the sporangium responsible for opening the sporangium and dispersing the spore; in mosses, thick-walled cells along the rim of the sporangium to which the peristome is attached

Anther—pollen-producing portion of the stamen

Antheridium—sperm-producing unicellular or multicellular organ of algae, fungi, bryophytes, and vascular plants other than angiosperms and gymnosperms

Apical dominance—suppression of growth in lateral buds by the release of auxin from the apical bud

Apical meristem—area of active cell division and growth at the tip of a root or shoot

Apomixis—reproduction without fusion of egg and sperm in an otherwise sexual structure or life cycle

Apoplastic pathway—movement or transport of materials between cell walls of adjacent cells and between the cell walls of a plant or organ

Ascocarp—multicellular structure of ascomycetes in which nuclear fusion and meiosis occur

Ascospore—spore produced by ascomycetes in the ascus

Ascus—specialized cell of ascomycetes in which haploid nuclei unite to form a zygote that undergoes meiosis to produce ascospores

Asexual reproduction—reproductive process, such as fragmentation, budding, or fission, that does not require union of gametes

Autotroph—organism capable of synthesizing required nutrients from inorganic substances in its environment

Axillary—buds or branches that grow in the upper angle between a leaf and the stem from which it grows; also called a lateral bud

Basidiocarp—multicellular structure of the basidiomycetes in which basidia are formed

Basidiospore—spore produced within the basidium of basidiomycetes

Basidium—reproductive cell of basidiomycetes in which nuclear fusion and meiosis occur, resulting in the formation of basidiospores

Bark—tissues outside the vascular cambium of woody stems

Berry—thin-skinned fruit that develops from a compound ovary and usually contains more than one seed

Binary fission—type of asexual reproduction in which cells divide into two equal parts but mitosis does not occur; characteristic of prokaryotes

Biogeochemical cycle—various nutrient cycles within ecosystems or on a global scale that involve both abiotic and biotic components

Biome—terrestrial community that extends over a large geographic area and is characterized by its climate and soils

Biotic—living elements in an environment

Blade—broad, expansive portion of a leaf; also called the *lamina*

Bract—modified, often small and reduced, leaflike structure

Bryophyte—nonseed, nonvascular plant (for example, moss, hornwort, or liverwort)

Bud—embryonic shoot containing meristem and often protected with covering young leaves; vegetative outgrowth from a yeast cell in a type of asexual reproduction

Bulb—short underground stem with enlarged and fleshy leaves containing stored food

Bundle sheath cells—layer of cells (parenchyma, sclerenchyma, or both) surrounding the vascular bundle of leaves; found in many plants but characteristic of 4-carbon (C_4) plants

Calvin-Benson cycle—light-independent chemical reactions of photosynthesis that occur in the stroma of the chloroplasts

Calyx—collection of sepals

Cambium—meristem that produces concentric rings of secondary xylem and phloem and consists of actively dividing cells

Capsule—dry fruit in angiosperms that splits open at maturity and develops from two or more carpels

Carnivore—animal that eats the flesh of another animal

Carpel—compartment of the pistil that contains an ovule

Carpellate—having one or more carpels but no functional stamens; also called *pistillate*

Casparian strip—band of suberin surrounding the walls of individual endodermal cells that blocks the flow of materials in the intercellular space, thereby blocking apoplastic transport

Cell membrane—outer boundary of a cell next to the cell wall that consists of a lipids, associated proteins, and carbohydrates; also called *plasma membrane*

Cellulose—insoluble polysaccharide consisting largely of glucose; a major component of cell walls in plants

Cell wall—rigid outermost boundary of plant cells, some bacteria, and protists

Centrioles—two structures found mainly in animal cells that consist of cylinders of microtubules

Centromere—area of the chromosome at which the spindle fibers attach

Chemiosmosis—harvesting of energy for phosphorylation of ADP to form ATP by use of an electrochemical gradient created by the movement of protons across a cell membrane

Chitin—structural polysaccharide or an amino sugar found in cell walls of fungi and exoskeletons of arthropods

Chlorophyll—green photosynthetic pigment of plant cells, some bacteria, and protists; it is found in chloroplasts and essential for photosynthesis

Chloroplasts—cell organelle in which photosynthetic pigments are contained; found in plants and algae

Chlorosis—loss or reduction in the amount of chlorophyll

Chromatid—one of the two strands of a duplicated chromosome, joined together by the centromere; it is visible in cells during cell division

Chromatin—dark-staining complex of DNA and proteins found in the nucleus; often visible during interphase of the cell cycle

Chromosome—linear sequence of DNA and protein found in cell nuclei; contains the genes of an organism

Chrysolaminarin—variant of the polymer laminarin and a principal storage material in diatoms

Climacteric—sharp rise in respiration during the initial stages of fruit ripening; thought to act as a trigger to promote ripening of fruit

Coenocytic—condition of fungi in which cells are multinucleate; the nuclei are not separated from one another by cell walls or membranes

Coenzyme—organic molecule that plays an accessory role in an enzyme-catalyzed reaction

Coleoptile—protective layer or sheath surrounding the emerging shoot of seedlings of the grass family

Coleorhiza—protective layer or sheath surrounding the emerging radicle of seedlings of the grass family

Collenchyma—supporting tissue or collection of cells; often found in regions of primary growth of stems and leaves

Community—two or more populations of different species interacting with one another and occupying the same area

Competition—interaction between two or more organisms attempting to obtain a resource that is required by both and is in limited supply

Competitive exclusion—situation in which one species eliminates the other through competition

Complete flower—flower having all flower parts: sepals, petals, stamens, and pistil

Compound leaf—leaf whose blade is divided into distinct smaller portions called *leaflets*

Conidium—asexual fungal spore that is not produced within a sporangium

Cork—secondary tissue produced by cork cambium

Cork cambium—secondary meristematic tissue found in outer region of stems and roots of angiosperms and gymnosperms

Corm—modified, vertical, underground stem in which food is accumulated in the form of starch

Corolla—collection of petals

Cortex—primary tissue region of a stem or root bounded on the inside by vascular tissue and on the outside by epidermis

Cotyledon—seed leaf within the embryo of angiosperms and gymnosperms; it serves as a food storage organ in dicots and a food-absorbing organ in monocots

Crassulacean acid metabolism (CAM)—type of C_4 photosynthesis in which phosphoenolpyruvate fixes CO_2 in C_4 compounds at night and the fixed CO_2 is transferred to RuBP of the Calvin-Benson cycle during the day

Cross-pollination—transfer of pollen from the anther of one plant to the stigma of another

Crossing over—exchange of genetic material between chromatids of homologous chromosomes during prophase I of meiosis

Cuticle—waxy or fatty layer on the outer walls of epidermal cells

Cytokinesis—separation of cytoplasm and cellular organelles into two daughter cells during telophase of cell division

Cytokinin—group of plant hormones that promote cell division and other metabolic effects

Cytoplasm—living substance of the cell excluding the nucleus; consists of water, dissolved materials (such as proteins, amino acids, carbohydrates, and lipids), and any cell organelles

Cytosol—semifluid portion of the cytoplasm

Day-neutral plant—plant that flowers independent of day length

Decomposers—organisms (bacteria, fungi, protists) within an ecosystem that feed on organic material, breaking it down and making it available for recycling

Dehydration synthesis—group of chemical reactions that join monomers to form polymers by removing the components that form water

Denaturation—process by which a protein becomes biologically inactive through loss of shape or conformation; it can occur from changes in pH, salinity, or temperature

Denitrification—form of bacterial metabolism in which nitrate is converted to gaseous nitrogen and released into the atmosphere

Deoxyribonucleic acid (DNA)—genetic material that directs protein synthesis in cells; capable of self-replication and determining the synthesis of RNA

Detritivore—consumer organism that obtains energy from dead or waste organic matter

Detritus—dead organisms and waste organic matter

Dicotyledon (dicot)—type of angiosperm having two cotyledons in its seed, net-veined leaves, and flower parts in groups of four or five

Diffusion—net movement of suspended or dissolved particles or molecules from a more concentrated region to a less concentrated region; the result is uniform distribution of the particles throughout the medium

Dihybrid cross—breeding event in which the parents differ in two traits

Dioecious—having unisexual flowers or cones; male flowers or cones are confined to one plant and female flowers or cones are confined to another plant of the same species

Diploid—having two sets of chromosomes in the nucleus

Dominant allele—allele that is fully expressed in the phenotype of a heterozygous cell

Double fertilization—unique characteristic of angiosperms in which one sperm fuses with an egg to form a diploid zygote and another sperm fuses with the polar nuclei of the embryo sac to form a triploid endosperm nucleus

Drupe—simple fleshy fruit containing one seed enclosed within a hard endocarp

Ecology—study of organisms, their interactions with each other, and their interaction with the physical environment

Ecosystem—biologic community and its associated abiotic environment

Elaioplast—plastid that stores lipid

Embryo—young plant of the sporophyte generation

Endergonic—nonspontaneous chemical reaction in which free energy is absorbed from surroundings

Endocarp—innermost layer of the fruit wall or pericarp

Endodermis—single layer of cells surrounding the vascular bundle (stele) in roots

Endoplasmic reticulum—system of interconnected cell membranes that form channels and compartments within the cytoplasm of the cell

Endospore—dormant bacterial structure consisting of one or a few cells with a tough outer wall; they are produced during periods of environmental or nutritional stress

Enzyme—protein that speeds the rate of a chemical reaction without being used up in the process

Epicotyl—upper portion of an embryo that extends above the point of attachment of the cotyledons

Epidermis—outermost layer of tissue that covers many plant parts and is usually one cell layer thick

Epigynous—flower structure in which the sepals, petals, and stamens appear to attach at the top of the ovary; also known as inferior ovary

Epiphyte—nonparasitic relationship between organisms in which one organism is attached to and grows on another organism

Ethylene—gaseous plant hormone that inhibits growth and promotes fruit ripening

Eukaryotic—pertaining to cells with distinct organelles and a nucleus with a membrane and chromosomes

Exergonic—pertaining to spontaneous chemical reactions in which there is a net release of free energy

Exocarp—outermost wall of the pericarp, or fruit wall

Fascicle—bundle of pine needles or other needle-like leaves of gymnosperms

Fermentation—type of respiration in which the hydrogens removed from the glucose molecule during glycolysis are returned to pyruvic acid to form products such as ethyl alcohol or lactic acid

Fertilization—union of egg and sperm to form a zygote

Filial—in genetics, the offspring of the parental generation; often designated F1 (first generation), F2 (second generation), and so on

Follicle—dry fruit derived from a single carpel; it opens along one side at maturity

Food chain—levels of feeding relationships among organisms in a community or ecosystem

Food web—complex, interconnected feeding relationships between all organisms in a community or ecosystem

Fragmentation—type of asexual reproduction in which the parent organism separates into smaller units, each of which develop into new adult organisms

Frond—fern leaf

Gametangium—cell, organ, or structure in which gametes are formed

Gametophyte generation—haploid, gamete-forming phase of the plant life cycle; see *alternation of generations*

Gemma—small clump of vegetative cells produced on the thallus of some plants (liverworts) and used as mode of asexual reproduction

Gene—discrete unit of hereditary information located on the chromosome and consisting of DNA

Generative cell—in angiosperms, the cell of the male gametophyte that divides to produce two sperm cells; in gymnosperms, the cell of the male gametophyte that divides to produce a sterile cell and a sperm cell

Genetics—science of heredity, or the study of inheritance

Genotype—genetic make-up of an organism

Geotropism—response of plant roots and shoots to the gravitational pull of the earth; also called *gravitropism*

Gibberellin—group of plant hormones responsible for many growth-promoting effects, particularly stem elongation

Glycolysis—anaerobic breakdown of glucose to form two molecules of pyruvic acid; it results in net production of two ATP molecules

Golgi apparatus—disk-shaped organelle that accumulates and packages materials synthesized by the cell; also called the *dictyosome*

Gram stain—dye developed by Danish bacteriologist Christian Gram; permits differentiation of bacteria into two classes based on the construction of their cell walls

Grana—series of stacked thylakoids within the chloroplast

Gravitropism—see *geotropism*

Ground meristem—area of active cell division that produces all the primary tissues except for the epidermis and vascular tissues

Guard cells—pairs of cells that control the opening and closing of the stomata

Gymnosperm—plant whose seeds are not enclosed within an ovary during their development

Haploid—having only one set of chromosomes

Haustoria—specialized fungal hyphae that penetrate individual cells of a host; common in parasitic fungi

Heartwood—portion of the secondary xylem in woody plants located near the center of the stem

Herbivore—animal that eats plants for food

Herbivory—consumption of plant material

Heterosporous—possessing two types of spores, a microspore and a megaspore

Heterotroph—organism that cannot manufacture organic compounds and therefore must consume other organisms to survive

Heterozygous—having two different alleles for the same trait

Homologous chromosomes—chromosomes that associate in pairs during the first stage of meiosis; they are generally the same size and shape

Homosporous—condition in which only one type of spore is produced

Homozygous—having two identical alleles for the same trait

Humus—decomposing organic matter in soil

Hybridization—creation of novel genetic combinations through selective breeding of organisms; often used to improve crop yield or disease resistance

Hydrologic cycle—cycling of water between the atmosphere and earth's surface

Hydrolysis—splitting of chemical bonds within a larger molecule by the addition of water

Hypha—single, threadlike filament of a fungus; collectively the hyphae form the mycelium

Hypocotyl—portion of the embryo or seedling between the point of attachment of the cotyledons and the radicle

Hypodermis—one or more layers of cells immediately beneath the epidermis and

distinct from the parenchyma cells of the cortex of certain plants

Hypogynous—flower structure in which the sepals, petals, and stamens are attached below the ovary; also known as superior ovary

Immediate ovary—see *perigynous*

Imperfect flower—flower lacking either stamens or pistil

Incomplete dominance—type of inheritance in which F1 hybrids have an appearance reflecting that of both parents

Indusium—membranous growth of the epidermis of a fern frond that covers and protects a sorus

Inferior ovary—see *epigynous*

Inflorescence—flower cluster with a definite arrangement of the individual flowers

Integument—outermost layer of the ovule that may develop into a seed coat

Intercalary meristem—center of cell division found in areas of the plant other than the apex

Internode—stem region between successive nodes

Isogamy—type of sexual reproduction in which the gametes or gametangia are of equal size; characteristic of some algae and fungi

Karyogamy—fusion of two nuclei

Karyokinesis—actual division of the cell nucleus

Krebs cycle—chemical cycle that completes the metabolic breakdown of glucose to carbon dioxide

Laminarin—polymer of glucose which serves as the principal storage material in the brown algae

Leaching—downward washing of nutrients or inorganic ions through soil due to the percolation of water

Leaf primordium—lateral outgrowth from the apical meristem that becomes a leaf

Leaf scar— mark left on a twig when a leaf separates due to abscission

Legume—member of the bean or pea family; type of dry fruit that splits open along both margins at maturity

Lenticel—slightly raised group of spongy cells in bark of woody stems that permit gas exchange between the interior of the plant and the atmosphere

Lichen—group of organisms that results from a parasitic relationship between members of the kingdoms Fungi and Protista

Lignin—abundant and important constituent of the secondary cell wall of many vascular plants

Linkage—situation in which genes are located on the same chromosome and therefore may be inherited together

Long-day plant—plant that is stimulated to flower when it is exposed to periods of daylight longer than some critical period

Macromolecule—large molecule of living matter formed by joining smaller molecules, usually by dehydration synthesis

Macronutrient—inorganic plant nutrients required in large quantities for proper growth and development

Megasporangium—structure in which megaspores are produced

Megaspore—spore that develops into the female gametophyte

Megasporocyte—diploid cell in which meiosis occurs, resulting in the production of four haploid megaspores; also called *megaspore parent cell*

Meiosis—process in which two subsequent nuclear divisions result in reduction of nuclear material from diploid to haploid

Meristem—region of cell division that produces undifferentiated cells

Mesocarp—middle layer of the pericarp

Micronutrient—inorganic nutrients required by plants in small or trace amounts

Microphyll—leaf having a single unbranched vein not associated with a leaf gap

Micropyle—opening in the integuments of an ovule through which the pollen tube gains access to the embryo sac or archegonium

Microspore—spore that develops into the male gametophyte

Microsporogenesis—meiotic division of the microspore mother cells to produce microspores

Middle lamella—layer of pectin material that glues adjacent plant cells together

Mitochondria—cell organelles containing the enzymes used in the Krebs cycle and electron transport chain

Mitosis—nuclear division in which chromosomes separate into two cells; each resulting daughter cell is genetically identical to the other and to the parent cell

Monocotyledon (monocot)—type of angiosperm having only one cotyledon in its seed, parallel-veined leaves, and flowers parts in groups of threes

Monohybrid cross—breeding event in which the parents differ by only one trait

Monomer—molecule that serves as a subunit for building a larger chain-like molecule known as a polymer

Multiple fleshy fruit—cluster of mature fruits produced by a cluster of flowers

Mutation—change in the DNA of a gene that creates genetic diversity

Mutualism—symbiotic relationship between two or more organisms in which all partners benefit from the association

Mycelium—mass or collection of fungal hyphae

Mycorrhizae—mutualistic association between a fungus and plant roots

Myxamoeba—haploid, flagellated, slime mold cells that emerge from spores and unite to form a zygote

Nitrification—type of bacterial metabolism in which ammonium is converted to nitrate

Node—region of a stem where leaves attach

Nucellus—protective ovule tissue surrounding a developing embryo sac

Nuclear membrane—porous membrane that encloses the nucleus of eukaryotic cells

Nucleolus—small, spherical body in the nucleus of cells that is the site of ribosomal RNA synthesis

Nucleotide—single unit or monomer of nucleic acid composed of sugar, phosphate, and a nitrogen base

Oogonium—female sex organ (often in algae and fungi) that contains one or several eggs and consists of a single cell

Organelle—specialized structure within a cell, such as the nucleus, mitochondria, or chloroplast

Osmosis—diffusion of water or any solvent across a differentially permeable membrane; in the absence of other forces, water will move from an area of higher concentration to one of lower concentration

Ovule—structure of a seed plant that contains the female gametophyte and can potentially develop into a seed

Oxidation—loss of an electron or hydrogen by an atom or molecule

Oxidative phosphorylation—formation of ATP from ADP and inorganic phosphate using energy derived from redox reactions of the electron transport chain

Palisade mesophyll—leaf tissue consisting of columnar, chloroplast-bearing, parenchyma cells located just beneath the epidermis

Parenchyma—tissue comprised of living cells of variable size and shape

Parthenocarpic—fruit development without fertilization that usually results in a seedless fruit

Peduncle—stalk of an inflorescence or single flower

Peptide bond—type of chemical bond formed between two amino acids of a polymer chain, formed by dehydration synthesis

Peptidoglycan—polymer found only in bacteria cell walls that consists of modified sugars cross-linked by short peptide chains

Perfect flower—flower with both stamens and pistils

Pericarp—fruit wall that develops from the ovary wall; it consists of three layers: exocarp, mesocarp, and endocarp

Pericycle—tissue found between the endodermis and phloem of a root

Periderm—outer bark of woody plants consisting mostly of cork cells

Perigynous—floral structure in which the sepals, petals, and stamens attach to the rim of a cuplike extension of the receptacle; also known as immediate ovary

Petal—usually conspicuously colored flower part important in attracting pollinators; collectively called the *corolla*

Petiole—leaf stalk

Phelloderm—tissue formed to the inside of the cork cambium

Phenotype—physical traits of an organism

Phloem—food-conducting vascular tissue of the plant; composed of sieve tube cells, parenchyma cells, and sclerenchyma cells

Photolysis—light-dependent splitting of water molecules that takes place in photosystem II of the light reactions of photosynthesis

Photoperiod—length or duration of daylight

Photorespiration—light-dependent production of glycolic acid in chloroplasts and its subsequent oxidation in peroxisomes

Photosynthesis—conversion of light, water, and carbon dioxide to carbohydrate that occurs in the chloroplast in the presence of chlorophyll; oxygen is released as a byproduct

Photosystem—aggregation of photosynthetic pigments and enzymes that operate as a functional unit

Phototropism—growth in which the direction of light is the determining factor; bending or turning of stems toward the light source

Phytochrome—light-absorbing pigment found in the cytoplasm of plants and some algae that is associated with processes such as flowering, seed germination, and dormancy

Pistillate—see *carpellate*

Pith—central tissue of dicot stems and some roots that consists of parenchyma cells

Plasma membrane—see *cell membrane*

Plasmid—small ring of DNA that carries additional genes separate from those of a bacterial chromosome

Plasmodesmata—strands of cytoplasm that extend between adjacent cells

Plasmodium—multinucleate, active form of a slime mold

Plasmogamy—fusion of cytoplasm of cells not accompanied by nuclear fusion

Plastid—cell organelle associated with the manufacture or storage of carbohydrate

Plumule—terminal bud of the embryo in a seed plant; portion of a young stem above the cotyledon(s)

Pneumatophore—extensions of roots of some trees growing in swampy habitats that function to aerate roots

Pollen—microspore of seed plants that contain the male gametophyte

Pollen tube—tube that forms on germination of pollen and carries the male gametes to the female gametophyte

Pollination—in angiosperms, transfer of pollen from the anther to the stigma of the pistil; in gymnosperms, transfer of pollen from the male cone to the female cone

Polymer—large chain-like molecule comprised of many monomers joined together

Polyploidy—chromosomal change in which the organism possesses more than two complete sets of chromosomes

Polysaccharide—carbohydrate polymer comprised of many monosaccharides joined together by dehydration synthesis

Population—group of individuals of the same species in a given geographic area

Primary cell wall—layer of cell wall deposited during the first period of cell growth and expansion

Primary meristem—tissue derived from the apical meristem and classified as protoderm, ground meristem, or procambium

Primary tissue—cells produced by a primary meristem or apical meristem

Procambium—primary meristem that produces primary xylem or phloem

Producer—organism that synthesizes food by photosynthesis

Prokaryotic—cell that lacks a distinct nucleus and other membrane-bound organelles

Propagule—unit of dispersal and reproduction, typically a spore or seed

Protein—organic compound comprised of carbon, hydrogen, oxygen, sulfur, and nitrogen

Prothallus—gametophyte generation of ferns and relatives

Protoderm—primary meristem that produces epidermis

Protonema—green, photosynthetic, threadlike or platelike growth from a moss spore or certain liverwort spores

Radicle—portion of the embryo within a seed that on germination develops into the primary root

Receptacle—expanded portion of the peduncle that serves as the point of attachment of flower parts

Recessive allele—allele that is completely masked in the phenotype of a heterozygous cell

Redox—pertaining to chemical reactions that involve the transfer of one or more electrons or hydrogens from one molecule to another, also called *reduction-oxidation reactions*

Reduction—gain of electrons or hydrogens by an atom or molecule; occurs in conjunction with oxidation

Renaturation—condition in which a protein regains is functional shape or conformation and thus can resume its biologic function

Rhizome—horizontal, underground stem that superficially appears like a root

Ribonucleic acid (RNA)—important cellular molecule that occurs in three forms and is involved in protein synthesis

Ribosome—small cellular particle composed of RNA and protein that is the site of protein synthesis

Root hairs—cellular extensions of the root epidermis that increase the surface area or absorption of water and nutrients from the soil

Root pressure—pressure that develops in roots due to osmosis and the activity of the endodermis

Saprobes—organisms that can obtain nutrients from dead organic matter

Sapwood—outer functional layers of xylem or wood in a tree trunk

Scarification—process of cutting or softening the seed coat in order to break seed dormancy

Sclerenchyma—dead tissue composed of cells with thick walls that strengthens and supports plant organs

Secondary cell wall—innermost layer of the cell wall that is formed in some cells after cell elongation has stopped

Secondary tissue—tissue derived from the lateral meristems (vascular cambium or cork cambium)

Seed—mature ovule containing an embryo, stored food, and a protective outer coat

Seed coat—outer layer of the seed that develops from integuments

Self-pollination—pollen transfer from male to female structures within the same plant, often within the same flower

Sepal—small, leaflike structure of the flower; collectively the sepals are called the *calyx*

Shoot—above-ground portion of a plant consisting of the stem, leaves, and flowers

Short-day plant—plant that flowers in response to exposure to a photoperiod less than some critical length

Species—group of organisms that have similar anatomic characteristics and can interbreed and produce fertile offspring

Spine—hard, sharp defensive structure; usually a modified leaf

Spongy mesophyll—tissue having loosely arranged cells with large air spaces; generally confined to the lower portion of the leaf cross-section

Sporangiophore—stalklike structure composed of hyphae that grow upright from the mycelium of certain fungi; these specialized hyphae produce globe-shaped sporangia at their tips

Sporangium—single-cell or multicellular structure in which spores are produced

Spore—single cell or aggregation of reproductive cells capable of developing directly into a gametophyte without fusion to another cell

Sporophyll—modified leaf that bears one or more sporangia

Sporophyte generation—diploid, spore-producing phase of the plant life cycle; see *alternation of generations*

Stamen—male or pollen-producing structure of the flower that consists of a stalk-like filament and a pollen-producing capsule called an *anther*

Staminate—flower with one or more stamens but no functional pistils

Statolith—gravity-sensitive body found within the cells of stems and roots

Stele—central core of a root or stem consisting of endodermis, pericycle, and primary xylem and phloem

Stem—portion of the plant that is above-ground and forms the body of the plant

Stolon—stem that grows horizontally on the ground surface

Stoma—small pores on leaves and stems of herbaceous plants that serve as pathways for gas exchange and transpiration

Stratification—process of exposing moist seeds to low temperatures in order to break seed dormancy

Strobilus—collection of sporophylls on a common axis, usually resembling a cone

Stroma—liquid contents of the chloroplast and the site of light-independent reactions of photosynthesis

Suberin—fatty substance (lipid) found in the cell walls of cork cells and the cells of the Casparian strip of the endodermis

Succession—process of biologic community development

Superior ovary—see *hypogynous*

Symbiotic—relating to two or more dissimilar organisms living together in a close association

Symplastic pathway—movement of materials between cells through the plasmodesmata

Synapsis—pairing of homologous chromosomes during prophase I of meiosis, a period during which crossing over may occur

Taxonomy—branch of biology concerned with classification of organisms

Thallus—multicellular, generally flattened plant body that is not organized into roots, shoots, stems, or leaves

Thigmotropism—response of a plant to touch or contact with a solid object

Thylakoid—sac formed from membranes and found in chloroplasts and cyanobacteria; thylakoids contain photosynthetic pigments

Translocation—movement or transport of food materials within a plant

Transpiration—loss of water vapor from plant parts, primarily through the stomata

Triglyceride—fat composed of three fatty acids bonded to a glycerol molecule

Trophic level—feeding level associated with the movement of energy through an ecosystem

Tropism—growth response of a plant to a directional external stimulus

Tube cell—cell within the male gametophyte (pollen grain) of seed plants that develops into the pollen tube

Tuber—swollen underground stem that is generally used for food storage

Turgor pressure—pressure within a cell due to movement of water into the cell

Vacuole—space within a cell filled with a watery fluid

Vascular bundle—strand of primary xylem and phloem surrounded by a layer of parenchyma cells known as the *bundle sheath*

Vascular cambium—cylindrical layer of meristematic cells that produce secondary xylem and phloem

Vesicle—small, membrane-bound sac within the cytoplasm of a cell

Xylem—complex tissue through which most of the water and dissolved minerals used by the plant are conducted

Zoospore—motile spore occurring in algae and fungi

Zygote—diploid cell resulting from the union on two gametes

Selected References

Brum, G.D., and McKane, L.K. *Biology: Exploring Life.* New York: John Wiley & Sons, 1989.

Edwards, G.I. *Biology the Easy Way*, 2nd ed. Woodbury, N.J.: Barron's, 1990.

Mauseth, J.D. *Botany: An Introduction to Plant Biology.* Philadelphia: W.B. Saunders Co., 1991.

Miller, Jr., G.T. *Living in the Environment: An Introduction to Environmental Science,* 7th ed. Belmont, Calif: Wadsworth Publishing, 1992.

Raven, P.H., Evert, R.Y., and Eichhorn, S.E. *Botany* (4th ed.). New York: Worth Publishers, 1986.

Stern, K.R. *Introductory Plant Biology*, 5th ed. Dubuque, Iowa: William C. Brown Publishers, 1991.

Wallace, R.A., King, J.L., and Sanders, G.P. *Biology: The Science of Life*, 3rd ed. Glenview, Ill.: Scott, Foresman & Co., 1990.

Index

i refers to an illustration; t, to a table

i refers to an illustration; t, to a table